Construction and Analysis of Cryptographic Functions

Lilya Budaghyan

Construction and Analysis of Cryptographic Functions

 Springer

Lilya Budaghyan
Department of Informatics
University of Bergen
Bergen
Norway

ISBN 978-3-319-36358-5   ISBN 978-3-319-12991-4
DOI 10.1007/978-3-319-12991-4

Springer Cham Heidelberg New York Dordrecht London
© Springer International Publishing Switzerland 2014
Softcover reprint of the hardcover 1st edition 2014

Printed on acid-free paper

Springer is part of Springer Science+Business Media (www.springer.com)

Preface

In this work we study discrete functions with optimal cryptographic properties such as APN, AB, bent and planar functions. We develop new methods for construction of these functions. In particular, we present the first infinite families of APN and AB functions CCZ-inequivalent to power functions which also served as a proof of existence of such AB functions. We construct 7 out of 11 known infinite families of quadratic APN functions CCZ-inequivalent to power functions, 4 of which are also AB when n is odd. Further, we construct new infinite families of planar functions over F_p^{2k}, where p is any odd prime, and we prove that these families of planar functions define a new family of commutative semifields of order p^{2k} when k is odd. After the works of Dickson (1906) and Albert (1952), this was the firstly found infinite family of commutative semifields which is defined for all odd primes p.

We investigate the relation between CCZ- and EA-equivalence (which are equivalence relations preserving the main cryptographic properties) for functions over finite fields and their important subfamilies (for example, bent, planar, Boolean functions, etc). We construct new classes of bent functions by applying CCZ-equivalence to non-bent vectorial functions. We further consider possible generalizations of CCZ-equivalence. We prove CCZ-inequivalence between some of the known power APN functions.

In 1998 Carlet, Charpin and Zinoviev characterized APN and AB functions via Boolean functions γ and proposed a problem to determine γ for the known APN and AB functions (these could potentially provide new bent functions and new invariants for APN and AB functions). We solve this problem for almost all known families of APN and AB functions.

In the present work we further solve a problem which dates back to 1974. In his thesis, Dillon introduced a family of bent functions denoted by H, where he was able to exhibit only functions belonging to the completed Maiorana-McFarland class. We prove that the class H contains functions which do not belong to the completed Maiorana-McFarland class.

We also study the relation between known classes of generalized bent functions given in trace representation and completed class of Maiorana-McFarland functions. Opposite to the binary case, we prove that there exist quadratic bent functions not belonging to the completed class of Maiorana-McFarland bent functions even when n

is even. Moreover, we prove that almost all of the known classes of generalized bent functions given in trace representation do not intersect with the completed Maiorana-McFarland class. This leads us to the conclusion that in general, the Maiorana-McFarland construction is less overall than in the binary case even for the case n even.

I would like to thank all the co-authors of the results presented in this work and, in particular, Claude Carlet, Tor Helleseth, Alexander Kholosha and Gregor Leander.

This research was supported by postdoctoral fellowships of Norwegian Research Counsel, Foundation Sciences Mathématiques de Paris, MIUR-Italy via PRIN 2006, also by Lavoro eseguito con il contributo della Provincia autonoma di Trento.

Contents

Chapter 1
Introduction

Discrete functions are functions from one finite set to another one and an important particular case of such functions is the case of Boolean functions, which are used at the heart of virtually all known digital systems: computers, telecommunications and cryptographic primitives, for example, all depend on the theory of Boolean functions. Moreover, functions with optimal cryptographic properties, define optimal objects in many domains of mathematics and information theory. Hence, construction and analysis of optimal cryptographic functions is connected with important mathematical problems and solution of these problems makes a valuable contribution to both mathematics and information theory.

Let p be a prime and n any positive integer. We denote by $\mathbb{F}_{p^n}$ the finite field with p^n elements and by $\mathbb{F}_p^n$ the n-dimensional vector space over $\mathbb{F}_p$. We study functions from $\mathbb{F}_p^n$ to $\mathbb{F}_p^m$, where n and m are arbitrary, which we call an (n, m, p)-function or a vectorial function. An $(n, m, 2)$-function we simply call an (n, m)-function or a vectorial Boolean function and an $(n, 1, 2)$-function is called a Boolean function in n variables.

In modern society, exchange and storage of information in an efficient, reliable and secure manner is of fundamental importance. Cryptographic primitives are used to protect information against eavesdropping, unauthorized changes and other misuse. In the case of symmetric cryptography ciphers are designed by appropriate composition of nonlinear Boolean functions. For example the security of block ciphers depends on S-boxes which are (n, m)-functions. For most of cryptographic attacks there is a certain property of functions which measures the resistance of the S-box to this attack. The two most powerful attacks on symmetric cryptosystems are differential and linear attacks and the respective cryptographic properties of functions are the nonlinearity and the differential uniformity. An interesting fact is that functions with optimal nonlinearity or differential uniformity define optimal objects in coding theory, finite geometry, sequence design, algebra, combinatorics, et al. Hence construction and analysis of these functions is important for all these theories and the present work is dedicated to this task.

There are many different Boolean functions even in very few variables so it is an important and very challenging task to find the best Boolean functions for each possible application. The abundance of Boolean functions means that it is usually

© Springer International Publishing Switzerland 2014
L. Budaghyan, *Construction and Analysis of Cryptographic Functions,*
DOI 10.1007/978-3-319-12991-4_1

impossible to solve these problems by using computer searches alone and one needs to invent more effective methods to construct new and better Boolean functions that can be efficiently implemented in electronic devices and thus save power and time in many applications.

An (n, m, p)-function F is called differentially δ-uniform if the equation $F(x) - F(x+a) = b$ has at most δ solutions for every nonzero element a of $\mathbb{F}_p^n$ and every b in $\mathbb{F}_p^m$. Functions with the smallest possible differential uniformity contribute an optimal resistance to the differential attack [5, 41]. In this sense differentially p^{n-m}-uniform functions, called perfect nonlinear (PN), are optimal. However when $p = 2$, PN functions exist only for n even and $m \leq n/2$. When $p = 2$ and $n = m$ differentially 2-uniform functions, called almost perfect nonlinear (APN), have the smallest possible differential uniformity.

The nonlinearity $\mathcal{NL}(F)$ of an (n, m)-function F is the minimum Hamming distance between all the component functions of F (that is, the functions $v \cdot F$ where "$\cdot$" denotes an inner product in $\mathbb{F}_2^m$ and all affine Boolean functions in n variables. The nonlinearity quantifies the level of resistance of the function to the linear attack: the higher is the nonlinearity $\mathcal{NL}(F)$ the better is the resistance of F to the linear attack [26, 40]. The functions achieving the upper bound on nonlinearity are called bent functions. All bent functions are also PN and vice versa, that is, these functions have optimal resistance against both linear and differential attacks. As mentioned above in the binary case PN (or bent) functions do not exist when $m = n$. In this case functions with the best possible nonlinearity are called maximally nonlinear. When n is odd maximally nonlinear functions are called almost bent (AB). When n is even the upper bound on the nonlinearity is still to be determined. All AB functions are APN, but the converse is not true in general. APN and AB functions define binary error correcting $[2^n, 2^n - 2n - 1, 5]$ codes optimal in certain sense [25]. Quadratic APN functions also define dual hyperovals [32, 48].

The nonlinearity and the differential uniformity (and, therefore, bentness, APN-ness and ABness), are invariant under affine, extended affine and CCZ-equivalences (in increasing order of generality). Two (n, m, p)-functions F and F' are called affine equivalent if one is equal to the other, composed on the left and on the right by affine permutations. They are called extended affine equivalent (EA-equivalent) if one is affine equivalent the other, added with an affine function. They are called Carlet-Charpin-Zinoviev equivalent (CCZ-equivalent) if their graphs $\{(x, F(x)) \mid x \in \mathbb{F}_p^n\}$ and $\{(x, F'(x)) \mid x \in \mathbb{F}_p^n\}$ are affine equivalent. CCZ-equivalence is the most general known equivalence relation of functions for which the nonlinearity and the differential uniformity are invariant.

The notion of CCZ-equivalence is difficult to handle, since checking whether two given functions are CCZ-equivalent or not is hard (at least when they share the same CCZ-invariant parameters). Building functions CCZ-equivalent (but not EA-equivalent) to a given function is hard too. The less general EA-equivalence is on the contrary simpler to check and, given some function, building EA-equivalent ones is very easy. Hence, identifying situations in which CCZ-equivalence reduces to EA-equivalence is useful. We investigate this problem and, in particular, we prove that this happens for all single output Boolean functions and that it does not, for functions from $\mathbb{F}_p^n$ to $\mathbb{F}_p^m$ under condition that m is greater or equal to the smallest divisor of n different from 1, which for n even case simply implies $m \geq 2$ [13, 15]. In [44] these results were extended to a more general framework of functions over finite abelian groups in which the condition on m is reduced to $m \geq 2$ also for n odd case and for p odd also includes $m = 1$.

We prove further that CCZ-equivalence coincides with EA-equivalence for all (single output or multi output) PN (or bent) functions, and, more generally, for all functions whose all derivatives are surjective [12, 16].

A question which has some importance for theoretical and practical reasons is whether CCZ-equivalence is really the most general equivalence relation of functions which is relevant to the block cipher framework. We show that trying to extend CCZ-equivalence to a more general notion in the same way as affine equivalence was extended to CCZ-equivalence (that is, by considering CCZ-equivalence of the indicators of the graphs of the functions instead of that of the functions themselves) leads in fact to the same CCZ-equivalence [13, 15].

Classification of bent, APN and AB functions is a hard open problem. Complete classification for APN and AB functions is known only for $n \leq 5$, [8]. For bent functions it can be done nowadays for $n \leq 8$, [37]. There are only a few classes of APN and AB functions known. Among them are six classes of APN power functions, four of which are also AB (for n odd). AB power functions are in one-to-one correspondence with sequences used for radars and for spread-spectrum communications. A binary sequence which can be generated by an LFSR is called an m-sequence or a maximum-length sequence if its period equals $2^n - 1$, which is the maximal possible value. Such a sequence can be used for radars and for code division multiple access (CDMA) in telecommunications, since it allows sending a signal which can be easily distinguished from any time-shifted version of itself. It is conjectured by Dobbertin that classification of APN and AB power functions is complete.

In this work we study the question of CCZ-equivalence for known power APN functions. We prove that two Gold functions x^{2^i+1} and x^{2^j+1} with $1 \leq i, j < n/2$, $i \neq j$, are CCZ-inequivalent, and that the Gold functions are CCZ-inequivalent to any Kasami and to Welch functions (except in particular cases) [17]. We also show that the inverse and Dobbertin APN functions are CCZ-inequivalent to each other and to all other known power APN mappings [17].

The other known families of APN and AB functions are a few classes of quadratic APN and AB functions constructed recently [7, 6, 11, 18–20]. Besides, for $6 \leq n \leq 9$ there is a large list of quadratic APN and AB functions for which infinite classes are still to be determined [9, 33, 49]. The only known example of APN functions CCZ-inequivalent to power functions and to quadratic functions is found in [8, 33] for $n = 6$ and construction of infinite families of such functions is still an open problem. Construction of infinite family of APN permutations over $\mathbb{F}_2^{2k}$ is an open problem too while an example of such a function is constructed for $n = 6$ in [10] by this disproving a longstanding conjecture about non-existence of such functions.

In this work we present seven out of eleven known infinite families of quadratic APN polynomials CCZ-inequivalent to power functions (four of which are also AB when n is odd) [11, 18–20]. First, we introduce two infinite classes of quadratic APN functions for n divisible by 3, respectively, 4, [18]. We prove that for n odd these functions are AB permutations. We show that, for $n \geq 12$, these functions are EA-inequivalent to power mappings and CCZ-inequivalent to Gold, Kasami, inverse and Dobbertin functions. This implies that for n even they are CCZ-inequivalent to all known APN functions. In particular, for $n = 12, 20, 24$, they are CCZ-inequivalent to any power mappings. These classes of binomials are the firstly found classes of APN functions CCZ-inequivalent to power mappings. Besides, they are the first counterexamples for the conjecture of [25] on nonexistence of quadratic AB functions inequivalent to the Gold maps. Further we discuss the possibility of generalization of the introduced APN binomials for other divisors of n.

An (n, n)-function F is called crooked if $F(x) + F(y) + F(z) + F(x + y + z) \neq 0$ for any three distinct elements x, y, z, $F(0) = 0$, and $F(x) + F(y) + F(z) + F(x + a) + F(y + a) + F(z + a) \neq 0$ for any $a \neq 0$ and x, y, z arbitrary [3]. On one hand, every crooked function gives rise to a distance regular rectagraph of diameter 3, and on the other hand every quadratic AB permutation is crooked [3]. The converse is not known, that is, whether a crooked function is necessarily a quadratic AB permutation. A rectagraph is a graph without triangles in which every pair of vertices at distance 2 lies in a unique 4-cycle. There are not too many constructions of rectagraphs known, especially rectagraphs of small diameter. Hence construction of such functions would provide not only interesting building blocks for symmetric cryptosystems but would also provide new distance regular rectagraphs. Nowadays only two families of crooked functions are known: one, the Gold functions, constructed in 1968 by Gold [34] in context of sequence design and rediscovered in 1993 by Nyberg [41], which gives Preparata graphs, and the other one is our family of binomials with n divisible by 3 [18].

Further we develop the method for constructing differentially 4-uniform quadratic polynomials introduced by Dillon [30] by proposing its various generalizations [11]. We construct a new infinite class of quadratic APN trinomials and a new potentially infinite class of quadratic APN hexanomials which we conjecture to be CCZ-inequivalent to power functions for $n \geq 6$ and we confirm this conjecture for $n \leq 10$ [11].

Then we present a method for constructing new quadratic APN functions from known ones [20]. Applying this method to the Gold power functions we construct an APN function $x^3 + \text{tr}_n(x^9)$ over $\mathbb{F}_{2^n}$. We prove that for $n \geq 7$ this function is CCZ-inequivalent to the Gold functions, and in the case $7 \leq n \leq 10$ it is CCZ-inequivalent to any power mapping and any APN polynomial belonging to the previously known families. This was the first APN polynomial CCZ-inequivalent to power functions whose all coefficients are in $\mathbb{F}_2$ and is still the only one which is defined for any n [20].

Further we give sufficient conditions on linear functions L_1 and L_2 from $\mathbb{F}_{2^n}$ to itself such that the function $L_1(x^3) + L_2(x^9)$ is APN over $\mathbb{F}_{2^n}$ [19]. We show that this can lead to many new cases of APN functions. In particular, we get two new families of APN functions $x^3 + a^{-1}\text{tr}_n^3(a^3 x^9 + a^6 x^{18})$ and $x^3 + a^{-1}\text{tr}_n^3(a^6 x^{18} + a^{12} x^{36})$ over $\mathbb{F}_{2^n}$ for any n divisible by 3 and $a \in \mathbb{F}_{2^n}^*$. We prove that for $n = 9$, these families are pairwise different and differ from all previously known families of APN functions, up to CCZ-equivalence. We also investigate further sufficient conditions under which the conditions on the linear functions L_1 and L_2 are satisfied.

For $p = 2$ bent (or PN) functions are the characteristic functions of elementary Hadamard difference sets [29]. Moreover they are employed to construct families of binary sequences, which possess properties well suited for application in code-division multiple-access communications systems [42]. Besides, some of these functions define o-polynomials (that is, permutations associated with hyperovals in a projective plane) [24]. For p odd PN functions were shown to be useful for constructing certain combinatorial objects such as partial difference sets, strongly regular graphs and association schemes [27, 45, 46].

Some of our results presented in this work are dedicated to analysis and construction of bent functions [12, 21–23]. As it was mentioned before, CCZ-equivalence of bent vectorial functions over $\mathbb{F}_2^n$ reduces to their EA-equivalence. We prove that in spite of this fact, CCZ-equivalence can be used for constructing bent functions which are new up to EA-equivalence and therefore to CCZ-equivalence: applying CCZ-equivalence to a non-bent vectorial Boolean function F which has some bent components, we get a function F' which also has some bent components and whose bent components are CCZ-inequivalent to the components of the original function F. Using this approach we construct classes of nonquadratic bent Boolean and bent vectorial functions [12].

Further we study the problem raised in [25]. In 1998, Carlet, Charpin and Zinoviev characterized APN and AB (n, n)-functions by means of associated $2n$-variable Boolean functions. In particular, they proved that a function F is AB if and only if the associated Boolean function γ_F is bent. This observation leads to potentially new bent functions associated to the known AB functions, or at least gives new insight on known bent functions. However, representations of γ_F were known only for Gold and inverse power APN functions and determining γ_F for the rest of AB and APN functions was an open problem. We determine the representation of γ_F for all known power AB functions and for almost all known families of APN polynomials. We also try to determine whether these bent functions (when F is AB) belong to the main known classes of bent functions [21]. In addition, CCZ-equivalence and EA-equivalence result in the equivalence of the associated functions γ_F. We note that if (n, n)-functions F and F' are CCZ-equivalent then there exists an affine permutation $\mathcal{L}$ of $\mathbb{F}_2^{2n}$ such that $\gamma_{F'} = \gamma_F \circ \mathcal{L}$. This implies that all affine invariants of the Boolean function γ_F (as weight, differential and linear properties, algebraic degree, et al.) can be used as CCZ-invariants for the (n, n)-function F. Hence, although studying γ_F is interesting by itself, there are also practical reasons for it: because they can be a source of potentially new bent functions when F is AB and because affine invariants of γ_F are CCZ-invariants for F.

In the present work we further solve a problem which dates back to 1974. In his thesis [29], Dillon introduced a family of bent functions denoted by H, where bentness is proven under some conditions which were not obvious to achieve. In this class, Dillon was able to exhibit only functions belonging to the completed Maiorana-McFarland class (the completed MM class), which is one of the largest and best known families of bent functions. Since then it was an open problem whether these two classes of bent functions differ. In [24] it was observed that the completed class of H contains all bent functions of the, so called, Niho type which were introduced in [31] by Dobbertin et al. We prove that two classes of binomial Niho bent functions do not belong to the completed MM class. This implies that the class H contains functions which do not belong to the completed Maiorana-McFarland class and, therefore, the class H is not contained in the completed MM class [23].

We also study the relation between known classes of generalized bent functions given in trace representation and completed class of Maiorana-McFarland functions. In the binary case, the completed MM class contains all quadratic bent functions which are the simplest and best understood. We prove that this does not hold in the generalized case. First, for p odd there exist quadratic bent functions over $\mathbb{F}_{p^n}$ when n is odd while Maiorana-McFarland bent functions are defined only for n even. For the case n even, we provide examples of quadratic generalized bent functions not belonging to the completed MM class. Moreover, we prove that almost all of the known classes of generalized bent functions given in trace representation do not intersect with the completed MM class. This leads us to the conclusion that in general, the Maiorana-McFarland construction is less overall than in the binary case even for the case n even [22].

When p is odd and $n = m$ PN functions are also called planar and they are in one-to-one correspondence with commutative semifields of odd order. A finite semfield is a ring with no zero-divisors, a multiplicative identity and left and right distributivity. Obviously every finite field is a semifield. The first finite semifields, different from finite fields, were the commutative semifields of Dickson [28] which have order p^{2k} with p an odd prime and k a positive integer. Dickson may have been led to study semifields following the publication of Wedderburn's Theorem [47], which appeared the year before [28] and which Dickson was the first person to provide a correct proof for (see Parshall [43]). As no new structures are obtained by removing commutativity, it is reasonable to investigate those structures which are non-associative instead. The role of semifields in projective geometry was confirmed following the introduction of coordinates in non-Desarguesian planes by Hall [35]. Subsequent to Hall's work, Lenz [38] developed and Barlotti [2] refined what is now known as the Lenz-Barlotti classification, under which semifields correspond to projective planes of Lenz-Barlotti type V.1. In some sense, modern interest in semifields can be traced back to the important work of Knuth [36]. In spite of considerable work in this field the only two previously known infinite families of commutative semifields (different from finite fields) of order p^n defined for any odd prime p were constructed by Dickson and Albert [1, 28].

In this work we present two infinite families of quadratic perfect nonlinear multinomials over $\mathbb{F}_{p^{2k}}$ where p is any odd prime and k a positive integer [14, 16]. We prove that in general these functions are CCZ-inequivalent to previously known PN mappings. Besides, we supply results indicating that the planar functions, we introduce, define new commutative semifields. After the works of Dickson (1906) and Albert (1952), these were the firstly found infinite families of commutative semifields which are defined for all odd primes p. One of the families, which we introduce, has been constructed by extension of a known family of APN functions over $\mathbb{F}_{2^{2k}}$ [6].

This shows that known classes of APN functions over fields of even characteristic can serve as a source for further constructions of PN mappings over fields of odd characteristics. This method, first introduced in [14] was further applied to the families of APN binomials from [18] to extend them to the families of planar binomials. Hence, results on classification of APN and PN functions have important consequences on classification of commutative semifields and projective planes.

Further we extend the family of PN functions constructed in [4, 39] to a larger (up to CCZ-equivalence) family of PN functions. This is done by using isotopisms of semifields (which are not strong). That is, extending the family of PN functions we still stay within the same family of commutative semifields (up to isotopic equivalence) [15].

The present work is organized as follows. In Chap. 2 we give general information which is necessary for presentation of the results. Chapter 3 is dedicated to the results on equivalence relations of functions. Bent functions are analyzed and constructed in Chap. 4. In Chap. 5 we construct infinite families of quadratic APN and AB functions. Chapter 6 presents infinite families of planar functions and corresponding commutative semifields.

References

1. A. A. Albert. On nonassociative division algebras. *Trans. Amer. Math. Soc.* 72, pp. 296–309, 1952.
2. A. Barlotti. Le possibili configurazioni del sistema delle coppie punto-retta (A; a) per cui un piano grafico risulta (A; a)-transitivo. *Boll. Un. Mat. Ital.* 12, 212–226, 1957.
3. T. Bending, D. Fon-Der-Flaass. Crooked functions, bent functions and distance-regular graphs. *Electron. J. Comb.*, 5 (R34), 14, 1998.
4. J. Bierbrauer. Commutative semifields from projection mappings. *Designs, Codes and Cryptography*, 61(2), pp. 187–196, 2011.
5. E. Biham and A. Shamir. Differential Cryptanalysis of DES-like Cryptosystems. *Journal of Cryptology* 4, no. 1, pp. 3–72, 1991.
6. C. Bracken, E. Byrne, N. Markin, G. McGuire. New families of quadratic almost perfect nonlinear trinomials and multinomials. Finite Fields and Their Applications **14**(3), pp. 703–714, 2008.
7. C. Bracken, E. Byrne, N. Markin, G. McGuire. A Few More Quadratic APN Functions. *Cryptography and Communications* 3(1), pp. 43–53, 2011.
8. M. Brinkman and G. Leander. On the classification of APN functions up to dimension five. *Proceedings of the International Workshop on Coding and Cryptography 2007* dedicated to the memory of Hans Dobbertin, pp. 39–48, Versailles, France, 2007.
9. K. A. Browning, J. F. Dillon, R. E. Kibler, M. T. McQuistan. APN Polynomials and Related Codes. *Journal of Combinatorics, Information and System Science*, Special Issue in honor of Prof. D.K Ray-Chaudhuri on the occasion of his 75th birthday, vol. 34, no. 1–4, pp. 135–159, 2009.
10. K. A. Browning, J. F. Dillon, M. T. McQuistan, A. J. Wolfe. An APN Permutation in Dimension Six. *Post-proceedings of the 9-th International Conference on Finite Fields and Their Applications Fq'09, Contemporary Math.*, AMS, v. 518, pp. 33–42, 2010.
11. L. Budaghyan and C. Carlet. Classes of Quadratic APN Trinomials and Hexanomials and Related Structures. *IEEE Trans. Inform. Theory*, vol. 54, no. 5, pp. 2354–2357, May 2008.

12. L. Budaghyan and C. Carlet. On CCZ-equivalence and its use in secondary constructions of bent functions. *Preproceedings of International Workshop on Coding and Cryptography WCC 2009*, pp. 19–36, 2009.

13. L. Budaghyan and C. Carlet. CCZ-equivalence of single and multi output Boolean functions. *Post-proceedings of the 9-th International Conference on Finite Fields and Their Applications Fq'09*, Contemporary Math., AMS, v. 518, pp. 43–54, 2010.

14. L. Budaghyan and T. Helleseth. New perfect nonlinear multinomials over $F_{p^{2k}}$ for any odd prime p. *Proceedings of the International Conference on Sequences and Their Applications SETA 2008*, Lecture Notes in Computer Science 5203, pp. 403–414, Lexington, USA, Sep. 2008.

15. L. Budaghyan and T. Helleseth. On Isotopisms of Commutative Presemifields and CCZ-Equivalence of Functions. Special Issue on Cryptography of *International Journal of Foundations of Computer Science*, v. 22(6), pp. 1243–1258, 2011. Preprint at http://eprint.iacr.org/2010/507

16. L. Budaghyan and T. Helleseth. New commutative semifields defined by new PN multinomials. *Cryptography and Communications: Discrete Structures, Boolean Functions and Sequences*, v. 3(1), pp. 1–16, 2011.

17. L. Budaghyan, C. Carlet, G. Leander. On inequivalence between known power APN functions. *Proceedings of the International Workshop on Boolean Functions: Cryptography and Applications, BFCA 2008*, Copenhagen, Denmark, May 2008.

18. L. Budaghyan, C. Carlet, G. Leander. Two classes of quadratic APN binomials inequivalent to power functions. *IEEE Trans. Inform. Theory*, 54(9), pp. 4218–4229, 2008.

19. L. Budaghyan, C. Carlet, G. Leander. On a construction of quadratic APN functions. *Proceedings of IEEE Information Theory Workshop, ITW'09*, pp. 374–378, Taormina, Sicily, Oct. 2009.

20. L. Budaghyan, C. Carlet, G. Leander. Constructing new APN functions from known ones. *Finite Fields and Their Applications*, v. 15, issue 2, pp. 150–159, April 2009.

21. L. Budaghyan, C. Carlet, T. Helleseth. On bent functions associated to AB functions. Proceedings of *IEEE Information Theory Workshop*, ITW'11, Paraty, Brazil, Oct. 2011.

22. L. Budaghyan, C. Carlet, T. Helleseth, A. Kholosha. Generalized Bent Functions and Their Relation to Maiorana-McFarland Class. *Proceedings of the IEEE International Symposium on Information Theory, ISIT 2012*, Cambridge, MA, USA, 1–6 July 2012.

23. L. Budaghyan, C. Carlet, T. Helleseth, A. Kholosha, S. Mesnager. Further Results on Niho Bent Functions. *IEEE Trans. Inform. Theory*, 58(11), pp. 6979–6985, 2012.

24. C. Carlet and S. Mesnager, "On Dillon's class H of bent functions, Niho bent functions and o-polynomials", *J. Combin. Theory Ser. A*, vol. 118, no. 8, pp. 2392–2410, Nov. 2011.

25. C. Carlet, P. Charpin and V. Zinoviev. Codes, bent functions and permutations suitable for DES-like cryptosystems. *Designs, Codes and Cryptography*, 15(2), pp. 125–156, 1998.

26. F. Chabaud and S. Vaudenay. Links between differential and linear cryptanalysis, *Advances in Cryptology—EUROCRYPT'94, LNCS*, Springer-Verlag, New York, 950, pp. 356–365, 1995.

27. Y. M. Chee, Y. Tan, and X. D. Zhang, "Strongly regular graphs constructed from p-ary bent functions", *J. Algebraic Combin.*, vol. 34, no. 2, pp. 251–266, Sep. 2011.

28. L. E. Dickson. On commutative linear algebras in which division is always uniquely possible. *Trans. Amer. Math. Soc* 7, pp. 514–522, 1906.

29. J. F. Dillon. Elementary Hadamard Difference sets. Ph. D. Thesis, Univ. of Maryland, 1974.

30. J. F. Dillon. APN Polynomials and Related Codes. *Polynomials over Finite Fields and Applications*, Banff International Research Station, Nov. 2006.

31. H. Dobbertin, G. Leander, A. Canteaut, C. Carlet, P. Felke, and P. Gaborit, "Construction of bent functions via Niho power functions," *J. Combin. Theory Ser. A*, vol. 113, no. 5, pp. 779–798, Jul. 2006.

32. Y. Edel. APN functions and dual hyperovals. *NATO Advanced Research Workshop*, 2008.

33. Y. Edel and A. Pott. A new almost perfect nonlinear function which is not quadratic. *Advances in Mathematics of Communications* 3, no. 1, pp. 59–81, 2009.

34. R. Gold. Maximal recursive sequences with 3-valued recursive crosscorrelation functions. *IEEE Trans. Inform. Theory*, 14, pp. 154–156, 1968.
35. M. Hall. Projective planes. *Trans. Amer. Math. Soc.* 54, pp. 229–277, 1943.
36. D. E. Knuth. Finite semifields and projective planes. *J. Algebra* 2, pp. 182–217, 1965.
37. P. Langevin, G. Leander. Counting all bent functions in dimension eight 99270589265934370305785861242880. *Des. Codes Cryptography* 59(1–3), pp. 193–205, 2011.
38. H. Lenz. Zur Begrundung der analytischen Geometrie. *S.-B. Math.-Nat. Kl. Bayer. Akad. Wiss.*, pp. 17–72, 1954.
39. G. Lunardon, G. Marino, O. Polverion, R. Trombetti. Symplectic spreads and quadric Veroneseans. Manuscript, 2009.
40. M. Matsui. Linear cryptanalysis method for DES cipher. *Advances in Cryptology-EUROCRYPT'93, LNCS*, Springer-Verlag, pp. 386–397, 1994.
41. K. Nyberg. Differentially uniform mappings for cryptography. *Advances in Cryptography, EUROCRYPT'93*, Lecture Notes in Computer Science 765, pp. 55–64, 1994.
42. J. D. Olsen, R. A. Scholtz, L. R. Welch. Bent-function sequences. *IEEE Trans. Inform. Theory* IT-28, pp. 858–864, 1982.
43. K. H. Parshall. In pursuit of the finite division algebra theorem and beyond. *Arch. Internat. Hist. Sci.* 33, pp. 274–299, 1983.
44. A. Pott, Y. Zhou. CCZ and EA equivalence between mappings over finite Abelian groups. *Des. Codes Cryptography* 66(1–3), pp. 99–109, 2013.
45. A. Pott, Y. Tan, T. Feng, and S. Ling. Association schemes arising from bent functions. *Des. Codes Cryptogr.*, vol. 59, no. 1–3, pp. 319–331, Apr. 2011.
46. Y. Tan, A. Pott, and T. Feng. Strongly regular graphs associated with ternary bent functions. *J. Combin. Theory Ser. A*, vol. 117, no. 6, pp. 668–682, Aug. 2010.
47. J. H. M. Wedderburn. A theorem on finite algebras. *Trans. Amer. Math. Soc.* 6, pp. 349–352, 1905.
48. S. Yoshiara. Notes on APN functions, semibiplanes and dimensional dual hyperovals. *Des. Codes Cryptography* 56(2–3), pp. 197–218, 2010.
49. Y. Yu, M. Wang, Y. Li. A matrix approach for constructing quadratic APN functions. *Pre-proceedings of the International Conference WCC 2013*, Bergen, Norway, 2013.

Chapter 2
Generalities

2.1 Equivalence Relations of Cryptographic Functions

Let p be a prime and n any positive integer. We denote by $\mathbb{F}_{p^n}$ the finite field with p^n elements and by $\mathbb{F}_p^n$ the n-dimensional vector space over $\mathbb{F}_p$. Further for any set E, we denote $E \setminus \{0\}$ by E^*.

In this work we study functions from $\mathbb{F}_p^n$ to $\mathbb{F}_p^m$ where n and m are arbitrary. Recall that a function F from $\mathbb{F}_p^n$ into $\mathbb{F}_p^m$ is called *an (n, m, p)-function* or *a vectorial function*. An $(n, m, 2)$-function we simply call *an (n, m)-function* or *a vectorial Boolean function* or *an S-box*. When $p = 2$ and $m = 1$ the function F is called *a Boolean function*. Opposite to vectorial functions which are denoted by capital letters, Boolean functions, and $(n, 1, p)$-functions in general, will be denoted by small letters. Clearly any (n, m, p)-function F can be presented in the form

$$F(x_1, ..., x_n) = (f_1(x_1, ..., x_n), ..., f_m(x_1, ..., x_n)),$$

where the $(n, 1, p)$-functions $f_1, ..., f_m$ are called the *coordinate functions* of the function F.

When $n = m$ it is often more convenient to identify the vector space $\mathbb{F}_p^n$ with $\mathbb{F}_{p^n}$ and consider functions from $\mathbb{F}_p^n$ to itself as mappings from $\mathbb{F}_{p^n}$ to itself. Any such function F has a unique representation as a univariate polynomial over $\mathbb{F}_{p^n}$ of degree smaller than p^n

$$F(x) = \sum_{i=0}^{p^n-1} c_i x^i, \quad c_i \in \mathbb{F}_{p^n}.$$

For any integer k, $0 \le k \le p^n - 1$, the number $w_p(k) = \sum_{s=0}^{n-1} k_s, 0 \le k_s \le p-1$, in the p-ary expansion $\sum_{s=0}^{n-1} p^s k_s$ of k is called the p-weight of k. The algebraic degree of a function $F : \mathbb{F}_{p^n} \to \mathbb{F}_{p^n}$ is equal to the maximum p-weight of the exponents of the monomials with nonzero coefficients in the polynomial $F(x)$:

$$d^\circ(F) = \max_{\substack{0 \le i \le p^n-1 \\ c_i \ne 0}} w_p(i).$$

© Springer International Publishing Switzerland 2014
L. Budaghyan, *Construction and Analysis of Cryptographic Functions*,
DOI 10.1007/978-3-319-12991-4_2

A function F from $\mathbb{F}_{p^n}$ to itself is

- *linear* if
$$F(x) = \sum_{0 \leq i < n} a_i x^{p^i}, \qquad a_i \in \mathbb{F}_{p^n};$$

- *affine* if F is a sum of a linear function and a constant;
- *Dembowski-Ostrom polynomial* (DO polynomial) if
$$F(x) = \sum_{0 \leq k \leq j < n} a_{kj} x^{p^k + p^j}, \quad a_{ij} \in \mathbb{F}_{p^n}; \tag{2.1}$$

- *quadratic* if it is a sum of a DO polynomial and an affine function.

If m is a positive divisor of n then a function F from $\mathbb{F}_{p^n}$ to $\mathbb{F}_{p^m}$ can be viewed as a function from $\mathbb{F}_{p^n}$ to itself and, therefore, it admits a univariate polynomial representation. More precisely, if $\mathrm{tr}_n^m(x)$ denotes the trace function from $\mathbb{F}_{p^n}$ into $\mathbb{F}_{p^m}$:

$$\mathrm{tr}_n^m(x) = x + x^{p^m} + x^{p^{2m}} + \dots + x^{p^{(n/m-1)m}},$$

(we shall write $\mathrm{tr}_n(x)$ instead of $\mathrm{tr}_n^1(x)$ when $m = 1$) then F can be represented in the form $\mathrm{tr}_n^m(\sum_{i=0}^{p^n - 1} c_i x^i)$. Indeed, there exists a function G from $\mathbb{F}_{p^n}$ to $\mathbb{F}_{p^n}$ (for example $G(x) = aF(x)$, where $a \in \mathbb{F}_{p^n}$ and $\mathrm{tr}_n^m(a) = 1$) such that F equals $\mathrm{tr}_n^m(G(x))$. Hence, any $(n, 1, p)$-function f can be written in a non-unique way as $\mathrm{tr}_n(G(x))$ where $G(x)$ is a polynomial over $\mathbb{F}_{2^n}$. Moreover, univariate representation of f can be written in the form of

$$f(x) = \sum_{j \in \Gamma_n} \mathrm{tr}_{o(j)}(a_j x^j),$$

where Γ_n is a set of integers obtained by choosing one element in each cyclotomic coset of p modulo $p^n - 1$, $o(j)$ is the size of the cyclotomic coset containing j and $a_j \in \mathbb{F}_{p^{o(j)}}$. This representation is unique up to the choice of cyclotomic coset representatives.

Let $n = 2k$ and f be an $(n, 1, p)$-function. Then f can have a bivariate representation defined as follows: we identify $\mathbb{F}_p^n$ with $\mathbb{F}_{p^k} \times \mathbb{F}_{p^k}$ and consider the argument of f as an ordered pair (x, y) of elements in $\mathbb{F}_{p^k}$. There exists a unique bivariate polynomial over $\mathbb{F}_{p^k}$ that represents f:

$$f(x) = \sum_{0 \leq i, j \leq p^k - 1} a_{i,j} x^i y^j.$$

Then the algebraic degree of f is equal to

$$d^\circ(f) = \max_{(i,j) \mid a_{i,j} \neq 0} (w_p(i) + w_p(j)).$$

And then the bivariate representation of f can be written in the form $f(x, y) = \mathrm{tr}_k(P(x, y))$, where $P(x, y)$ is some polynomial of two variables over $\mathbb{F}_{p^k}$.

When m is not a divisor of n, the univariate representation of an (n, m, p)-function F in the field is not convenient. We need then to see F as a function from $\mathbb{F}_p^n$ to $\mathbb{F}_p^m$ and the natural way of representing it is by its *algebraic normal form ANF*:

$$F(x) = \sum_{u \in \mathbb{F}_p^n} a_u \prod_{i=1}^{n} x_i^{u_i}, \quad a_u \in \mathbb{F}_p^m,$$

(this sum being calculated in $\mathbb{F}_p^m$). The way to obtain one representation of a function from the other is recalled in [38] for the binary case. The algebraic degree $d^\circ(F)$ of F equals the degree of its ANF. The minimum algebraic degree of all nonzero linear combinations of the coordinate functions of F is called the *minimum degree* of the function F and is denoted by $\min d^\circ(F)$.

A function F from $\mathbb{F}_p^n$ to $\mathbb{F}_p^m$, where n and m are arbitrary, is called *balanced* if every element of $\mathbb{F}_p^m$ has the same number p^{n-m} of pre-images. Balanced functions from $\mathbb{F}_p^n$ to itself are permutations of $\mathbb{F}_p^n$. In some cases properties of vectorial functions can be described by similar properties of its component functions. For instance, an (n, m)-function F is balanced if and only if all nonzero linear combinations of the coordinate functions of F are balanced, that is if and only if the Boolean function $c \cdot F$ is balanced for every nonzero $c \in \mathbb{F}_2^m$, where "$\cdot$" denotes the usual inner product in $\mathbb{F}_2^m$ (see [35]).

Let n, m and δ be any positive integers. A function F from $\mathbb{F}_p^n$ to $\mathbb{F}_p^m$ is called *differentially δ-uniform* if all the equations

$$F(x + a) - F(x) = b, \qquad a \in \mathbb{F}_p^{n*}, \ b \in \mathbb{F}_p^m, \qquad (2.2)$$

have at most δ solutions. Differential uniformity measures the resistance of a function, used as an S-box in a cryptosystem, to differential attack: the smaller is differential uniformity the better is the resistance [7, 97]. In this sense differentially p^{n-m}-uniform functions, called *perfect nonlinear* (PN), are optimal. Clearly, a function F is perfect nonlinear if and only if for any $a \in \mathbb{F}_p^{n*}$ the function $D_a F(x) = F(x+a) - F(x)$, called the *derivative of F in the direction of a*, is balanced. When $n = m$ this condition implies that all derivatives of F in non-zero directions are permutations, and in this case F is also called *a planar function*. Planar functions were introduced in 1968 by Dembowski and Ostrom [49] in the context of finite geometry to describe projective planes with specific properties. It is obvious that planar functions exist only for p odd since if p is even and x_0 is a solution of (2.2) then $x_0 + a$ is a solution too. The functions from $\mathbb{F}_2^n$ to itself, whose derivatives $D_a F$, $a \in \mathbb{F}_2^{n*}$, are 2-to-1 mappings, possess the best possible resistance to differential cryptanalysis and are called *almost perfect nonlinear* (APN).

There are several equivalence relations of functions for which differential uniformity is an invariant. Due to these equivalence relations, having only one PN (or APN) function, one can generate a huge class of PN (resp. APN) functions. The terminology for these equivalence relations was introduced in [24] while the ideas behind this terminology go back to the works of Nyberg [98] and Carlet, Charpin and Zinoviev [38].

Definitions for equivalences below are given for functions from $\mathbb{F}_p^n$ to $\mathbb{F}_p^m$. However they can be naturally extended to functions from A to B where A and B are arbitrary groups [24, 101]. Two functions F and F' from $\mathbb{F}_p^n$ to $\mathbb{F}_p^m$ are called

- *affine equivalent* (or *linear equivalent*) if $F' = A_1 \circ F \circ A_2$, where the mappings A_1 and A_2 are affine (resp. linear) permutations of $\mathbb{F}_p^m$ and $\mathbb{F}_p^n$, respectively;
- *extended affine equivalent* (EA-equivalent) if $F' = A_1 \circ F \circ A_2 + A$, where the mappings $A : \mathbb{F}_p^n \to \mathbb{F}_p^m$, $A_1 : \mathbb{F}_p^m \to \mathbb{F}_p^m$, $A_2 : \mathbb{F}_p^n \to \mathbb{F}_p^n$ are affine, and where A_1, A_2 are permutations;
- *Carlet-Charpin-Zinoviev equivalent* (CCZ-equivalent) if for some affine permutation $\mathcal{L}$ of $\mathbb{F}_p^n \times \mathbb{F}_p^m$ the image of the graph of F is the graph of F', that is, $\mathcal{L}(G_F) = G_{F'}$ where $G_F = \{(x, F(x)) \mid x \in \mathbb{F}_p^n\}$ and $G_{F'} = \{(x, F'(x)) \mid x \in \mathbb{F}_p^n\}$.

Although different, these equivalence relations are connected to each other. It is obvious that linear equivalence is a particular case of affine equivalence, and that affine equivalence is a particular case of EA-equivalence. As shown in [38], EA-equivalence is a particular case of CCZ-equivalence and every permutation is CCZ-equivalent to its inverse. The algebraic degree of a function (if it is not affine) is invariant under EA-equivalence but, in general, it is not preserved by CCZ-equivalence. Let us recall why the structure of CCZ-equivalence implies this: for a function F from $\mathbb{F}_p^n$ to $\mathbb{F}_p^m$ and an affine permutation

$$\mathcal{L}(x, y) = (L_1(x, y), L_2(x, y))$$

of $\mathbb{F}_p^n \times \mathbb{F}_p^m$, where $L_1 : \mathbb{F}_p^n \times \mathbb{F}_p^m \to \mathbb{F}_p^n$ and $L_2 : \mathbb{F}_p^n \times \mathbb{F}_p^m \to \mathbb{F}_p^m$, we have

$$\mathcal{L}(G_F) = \{(F_1(x), F_2(x)) : x \in \mathbb{F}_p^n\}$$

where

$$F_1(x) = L_1(x, F(x)),$$

$$F_2(x) = L_2(x, F(x)).$$

$\mathcal{L}(G_F)$ is the graph of a function if and only if the function F_1 is a permutation. The function CCZ-equivalent to F whose graph equals $\mathcal{L}(G_F)$ is then

$$F' = F_2 \circ F_1^{-1}.$$

The composition by the inverse of F_1 modifies in general the algebraic degree, except, for instance, when $L_1(x, y)$ depends only on x, which corresponds to EA-equivalence of F and F' (see Proposition 1 below proven in [24] for $p = 2$ and $n = m$ and whose proof obviously extends to the general case).

Proposition 1 [24] *Let F and F' be two (n, m, p)-functions where p is a prime and n and n are any positive integers. The function F' is EA-equivalent to the function F or to the inverse of F (if it exists) if and only if there exists a linear permutation $\mathcal{L} = (L_1, L_2)$ on $\mathbb{F}_2^n \times \mathbb{F}_2^m$ such that $\mathcal{L}(G_F) = G_{F'}$ and the function L_1 depends only on one variable, i.e. $L_1(x, y) = L(x)$ or $L_1(x, y) = L(y)$.*

For quite a long time it was believed that CCZ-equivalence class of an arbitrary function F can be completely described by means of EA-equivalence and the inverse F (if F is a permutation). In [17, 24], it is proven to be false: CCZ-equivalence is much more general. However, there are particular cases of functions for which CCZ-equivalence can be reduced to EA-equivalence. For instance, CCZ-equivalence coincides with

- EA-equivalence for planar functions [21, 23];
- linear equivalence for DO planar functions [21, 23];
- EA-equivalence for all functions whose derivatives are surjective [23];
- EA-equivalence for all Boolean functions [20];
- EA-equivalence for all vectorial bent Boolean functions [19];
- EA-equivalence for two quadratic APN functions (conjectured by Edel, proven by Yoshiara [109]).

It is useful to know cases where CCZ- and EA-equivalences coincide because in general it is very difficult to determine whether two functions are CCZ-equivalent or not while EA-equivalence is much simpler and has a nice invariant, algebraic degree of a function. Besides, if the minimum degree of F is grater than 1 then the minimum degree is also EA-invariant. Obviously, the algebraic and minimum degrees of a function are not invariant under the inverse transformation. For some CCZ-invariants one can see [64], for instance.

Nowadays, CCZ-equivalence is the most general known equivalence relation of functions preserving PN and APN properties and it is appealing to find a more general equivalence for which PN and APN properties are invariants. We make an attempt to solve this problem by studying the indicators of the graphs of functions. For a given function F from $\mathbb{F}_p^n$ to $\mathbb{F}_p^m$, let us denote the indicator of its graph G_F by 1_{G_F}, that is,

$$1_{G_F}(x, y) = \begin{cases} 1 & \text{if} \quad y = F(x) \\ 0 & \text{otherwise} \end{cases}.$$

However, as we shall prove further, two (n, m, p)-functions F and F' are CCZ-equivalent if and only if their indicators 1_{G_F} and $1_{G_{F'}}$ are CCZ-equivalent (see Theorem 7 or [20, 22]).

2.2 Bent Functions

Boolean bent functions were first introduced by Rothaus in 1976 as an interesting combinatorial object with the important property of having the maximum Hamming distance to the set of all affine functions. Later the research in this area was stimulated by the significant relation to the following topics in computer science: coding theory, sequences and cryptography (design of stream ciphers and S-boxes for block ciphers). Kumar, Scholtz and Welch in [84] generalized the notion of Boolean bent functions to the case of functions over an arbitrary finite field. Complete classification of bent functions looks hopeless even in the binary case. In the case of generalized bent functions, things are naturally much more complicated. However, many explicit methods are known for constructing bent functions either from scratch or based on other, simpler bent functions.

2.2.1 The Case of Even Characteristic

Let f be a Boolean function over $\mathbb{F}_2^n$. The *Hamming weight wt(f)* of the function f is the size of its *support* $\{x \in \mathbb{F}_2^n : f(x) \neq 0\}$. The *Hamming distance $d(f, g)$* between two Boolean functions f and g is the size of the set $\{x \in \mathbb{F}_2^n : f(x) \neq g(x)\}$. The minimum distance $\mathcal{NL}(f)$ between f and all affine Boolean functions is called the *nonlinearity* of the Boolean function f.

The *nonlinearity* of an (n, m)-function F is the minimum Hamming distance between all nonzero linear combinations of the coordinate functions of F and all affine Boolean functions on n variables. Clearly the nonlinearity of F is described by the nonlinearities of the Boolean functions $b \cdot F$

$$\mathcal{NL}(F) = \min_{b \in \mathbb{F}_2^m, b \neq 0} \mathcal{NL}(b \cdot F),$$

The linear cryptanalysis, introduced by Matsui [93], is based on finding affine approximations to the action of a cipher, therefore the linear attack on a function F is successful if $\mathcal{NL}(F)$ is small.

If we consider a Boolean function as valued in $\{0, 1\} \subset \mathbb{Z}$ then the nonlinearity can be described by Walsh transform. Let F be an (n, m)-function. The function $\lambda_F : \mathbb{F}_2^n \times \mathbb{F}_2^m \to \mathbb{Z}$ defined by

$$\lambda_F(a, b) = \sum_{x \in \mathbb{F}_2^m} (-1)^{b \cdot F(x) + a \cdot x}, \quad a \in \mathbb{F}_2^n, \quad b \in \mathbb{F}_2^m,$$

is called the *Walsh transform* of the function F. For any elements $a \in \mathbb{F}_2^n$ and $b \in \mathbb{F}_2^m$ the value $\lambda_F(a, b)$ is called the *Walsh coefficient* of F and the set

$$\Lambda_F = \{\lambda_F(a, b) : a \in \mathbb{F}_2^n, b \in \mathbb{F}_2^{m*}\}$$

is called the *Walsh spectrum* of F. The set

$$\Lambda'_F = \{|\lambda_F(a, b)| : a \in \mathbb{F}_2^n, b \in \mathbb{F}_2^{m*}\}$$

is called the *extended Walsh spectrum* of F. We also denote

$$\lambda_F = \max_{a \in \mathbb{F}_2^n, b \in \mathbb{F}_2^{m*}} |\lambda_F(a, b)|.$$

In case of a Boolean function f the Walsh transform is simply defined as

$$\lambda_f(a) = \sum_{x \in \mathbb{F}_2^n} (-1)^{f(x)+a \cdot x}, \quad a \in \mathbb{F}_2^n.$$

The Walsh transform of a function does not depend on a particular choice of the inner product in $\mathbb{F}_2^n$. If we identify $\mathbb{F}_2^n$ with $\mathbb{F}_{2^n}$ then we can take $x \cdot y = \mathrm{tr}_n(xy)$ and the Walsh transform of an (n, m)-function F can be defined as

$$\lambda_F(a, b) = \sum_{x \in \mathbb{F}_{2^n}} (-1)^{\mathrm{tr}_m(bF(x))+\mathrm{tr}_n(ax)}, \quad a \in \mathbb{F}_{2^n}, b \in \mathbb{F}_{2^m}.$$

If n is even then a Boolean function f can be considered as $f : \mathbb{F}_{2^{n/2}} \times \mathbb{F}_{2^{n/2}} \to \mathbb{F}_2$. In this case, we can take $(x, y) \cdot (x', y') = \mathrm{tr}_{n/2}(xx' + yy')$ where $\mathrm{tr}_{n/2}(x)$ is the trace function over $\mathbb{F}_{2^{n/2}}$. Then the Walsh transform of f is the function:

$$\lambda_f(a, a') = \sum_{x,y \in \mathbb{F}_{2^{n/2}}} (-1)^{f(x,y)+\mathrm{tr}_{n/2}(ax+a'y)}, \quad a, a' \in \mathbb{F}_{2^{n/2}}.$$

One can easily note that for any (n, m)-function F and any elements $a \in \mathbb{F}_2^n$, $b \in \mathbb{F}_2^m$ we have

$$\lambda_F(a, b) = 2^n - 2wt(b \cdot F(x) + a \cdot x) = 2^n - 2d(b \cdot F(x), a \cdot x).$$

Then

$$d(b \cdot F(x), a \cdot x) = 2^{n-1} - \frac{1}{2}\lambda_F(a, b),$$

$$d(b \cdot F(x), a \cdot x + 1) = 2^{n-1} + \frac{1}{2}\lambda_F(a, b).$$

This gives the connection between the nonlinearity of F and the values of its Walsh transform

$$\mathcal{NL}(F) = 2^{n-1} - \frac{1}{2}\lambda_F.$$

The nonlinearity and the extended Walsh spectrum of an (n, m)-function are invariant under CCZ-equivalence [38].

It can be easily proved that the Walsh transform of any Boolean function f in n variables satisfies Parseval's relation

$$\sum_{a \in \mathbb{F}_2^n} \lambda_F(a)^2 = 2^{2n}. \tag{2.3}$$

Parseval's relation makes clear that the nonlinearity $\mathcal{NL}(F)$ of any (n, m)-function F has an upper bound

$$\mathcal{NL}(F) \le 2^{n-1} - 2^{\frac{n}{2}-1}.$$

This bound is called the *universal bound*. Functions achieving this bound have the optimal nonlinearity and they are called *bent*. Bent functions exist only for n even and $m \leq n/2$ (see [99]). A function $F : \mathbb{F}_2^n \to \mathbb{F}_2^m$ is bent if and only if one of the following conditions holds (see [35]):

(i)   for any nonzero $c \in \mathbb{F}_2^m$ the Boolean function $c \cdot F$ is bent;
(ii) $\lambda_F(a, b) = \pm 2^{\frac{n}{2}}$ for any $a \in \mathbb{F}_2^n, b \in \mathbb{F}_2^{m*}$.
(iii) F is PN.

The algebraic degree of a bent Boolean function in $n > 2$ variables is at most $\frac{n}{2}$, [103]. If f is a bent Boolean function in n variables then its dual f' is the Boolean function defined by

$$\lambda_f(a) = 2^{\frac{n}{2}}(-1)^{f'(a)}.$$

Obviously, f' is also bent and its dual is f itself.

There are many known constructions of bent functions, see [34, 35] for most of them. However, complete classification of bent functions seems hopeless, it is an open problem for $n \geq 10$ (see [86] for classification with $n = 8$). The main known classes of bent functions are the Maiorana-McFarland class and the PS_{ap} class. An n-variable Boolean bent function belongs to the *Maiorana-McFarland class (MM class)* if, writing its input in the form (x, y), with $x, y \in \mathbb{F}_2^{n/2}$, the corresponding output equals $x \cdot \pi(y) + g(y)$, where π is a permutation of $\mathbb{F}_2^{n/2}$ and g is a Boolean function over $\mathbb{F}_2^{n/2}$. The *completed class of Maiorana-McFarland's functions (completed MM class)* is the set of all functions which are EA-equivalent to Maiorana-McFarland functions. In general, for any set S of functions we call the set S' of all functions EA-equivalent to the functions in S *the completed class of S*. The completed MM class contains all quadratic bent Boolean functions [54].

A bent Boolean function belongs to PS_{ap} if it has the form $f(x, y) = g\left(\frac{x}{y}\right)$ where g is a balanced Boolean function on $\mathbb{F}_{2^{n/2}}$ which vanishes at 0 (with the convention $\frac{1}{0} = 0$). These functions have the peculiarity that their algebraic degree equals $n/2$ (i.e. is optimal), which allows in some cases to exclude that a given bent function belongs to the completed class of PS_{ap}.

There are also a few classes of bent Boolean functions known in trace representation, in particular power bent functions (which can also be called monomial functions), that is, functions of the form $\mathrm{tr}_n(ax^i)$ with the following exponents:

- the Gold exponents $i = 2^j + 1$, where $n/\gcd(j, n)$ is even and $a \notin \{b^i, b \in \mathbb{F}_2^n\}$ (these functions belong to the Maiorana-McFarland class);
- the Dillon exponents of the form $i = j(2^{n/2} - 1)$, where $\gcd(j, 2^{n/2} + 1) = 1$ and $a \in \mathbb{F}_{2^{n/2}}$ is such that $\sum_{x \in \mathbb{F}_{2^{n/2}}} (-1)^{\mathrm{tr}_{n/2}(1/x + ax)} = 0$ with $1/0 = 0$ (these functions belong to the PSap class) [53];
- the Kasami exponents $i = 2^{2j} - 2^j + 1$, where $\gcd(j, n) = 1$ and $a \notin \{b^3, b \in \mathbb{F}_{2^n}\}$ (see [55, 87]);
- the exponent $i = (2^{n/4} + 1)^2$ where n is divisible by 4 but not by 8 and $a = a'b^i$ with $a' \in w\mathbb{F}_{2^{n/4}}$, $w \in \mathbb{F}_4 \setminus \mathbb{F}_2$, $b \in \mathbb{F}_{2^n}$ (this function belongs to the Maiorana-McFarland class) [41, 87];

- the exponent $i = 2^{n/3} + 2^{n/6} + 1$, where n is divisible by 6 and $a = a'b^i$ with $a' \in \mathbb{F}_{2^{n/2}}$ such that $\text{tr}_{n/2}^{n/6}(a') = 0$, $b \in \mathbb{F}_{2^n}$ (this function belongs to the Maiorana-McFarland class) [33].

A still simpler bent function (but which is not expressed by means of the function tr_n itself) is $\text{tr}_{n/2}(x^{2^{n/2+1}})$, which belongs to MM class.

Another interesting case of bent Boolean functions are so-called Niho bent functions. Recall that a positive integer d (always understood modulo $2^n - 1$) is said to be a *Niho exponent* and x^d is a *Niho power function* if the restriction of x^d to $\mathbb{F}_{2^m}$, for $m = n/2$, is linear or, in other words, $d \equiv 2^j \pmod{2^m - 1}$ for some $j < n$. As we consider $\text{tr}_n(ax^d)$ with $a \in \mathbb{F}_{2^n}$, without loss of generality, we can assume that d is in the normal form, i.e., with $j = 0$. Then we have a unique representation $d = (2^m - 1)s + 1$ with $2 \le s \le 2^m$. The simplest example of an infinite class of Niho bent functions is the quadratic function $\text{tr}_m(ax^{2^m+1})$ with $a \in \mathbb{F}_{2^n}^*$. In this case, $s = 1/2$ (interpret $1/2$ as an inverse of 2 modulo $2^m + 1$) and $2d = 2^m + 1$. Other known classes are:

(1) Two functions from [62] that are binomials of the form

$$f(x) = \text{tr}_n(\alpha_1 x^{d_1} + \alpha_2 x^{d_2}), \tag{2.4}$$

where

$$2d_1 = 2^m + 1 \in \mathbb{Z}/(2^n - 1)\mathbb{Z}$$

and $\alpha_1, \alpha_2 \in \mathbb{F}_{2^n}^*$ are such that $(\alpha_1 + \alpha_1^{2^m})^2 = \alpha_2^{2^m+1}$. Equivalently, denoting $a = (\alpha_1 + \alpha_1^{2^m})^2$ and $b = \alpha_2$ we have $a = b^{2^m+1} \in \mathbb{F}_{2^m}^*$ and

$$f(x) = \text{tr}_m(ax^{2^m+1}) + \text{tr}_n(b\, x^{d_2}).$$

Note that if $b = 0$ and $a \neq 0$ then f is also bent but becomes quadratic equal to the function mentioned above. The possible values of d_2 are:
$d_2 = (2^m - 1)3 + 1$,
$6d_2 = (2^m - 1) + 6$ (with the condition that m is even).
These functions have algebraic degree m.

(2) A function from [62, 88] which has the form

$$\text{tr}_n\left(ax^{2^m+1} + \sum_{i=1}^{2^{r-1}-1} x^{(2^m-1)\frac{i}{2^r}+1}\right) \tag{2.5}$$

with $r > 1$ satisfying $\gcd(r, m) = 1$ and $a \in \mathbb{F}_{2^n}$ is such that $a + a^{2^m} = 1$. This function belongs to the completed Maiorana-McFarland class.

(3) A few other functions found recently in [37], which are given in bivariate form.

The first class of binomial Niho bent functions was extended in [77] by removing the restriction on coefficient b.

We shall also analyze in this work the so-called class H of bent Boolean functions which was introduced by Dillon in his thesis [54]. The functions $f : \mathbb{F}_2^n \to \mathbb{F}_2$ in this class are defined in their bivariate form as

$$f(x, y) = \mathrm{tr}_m\left(y + xF(yx^{2^m-2})\right), \tag{2.6}$$

where $x, y \in \mathbb{F}_{2^m}$, $n = 2m$ and F is a permutation of $\mathbb{F}_{2^m}$ such that $F(x) + x$ does not vanish and for any $\beta \in \mathbb{F}_{2^m}^*$, the function $F(x) + \beta x$ is 2-to-1 (i.e., the pre-image of any element of $\mathbb{F}_{2^m}$ is either a pair or the empty set).

As observed by Carlet and Mesnager [37], this class can be slightly extended into a class $\mathcal{H}$ defined as the set of (bent) functions g satisfying

$$g(x, y) = \begin{cases} \mathrm{tr}_m\left(xG\left(\frac{y}{x}\right)\right) & \text{if} \quad x \neq 0 \\ \mathrm{tr}_m(\mu y) & \text{if} \quad x = 0 \end{cases}, \tag{2.7}$$

where $\mu \in \mathbb{F}_{2^m}$ and G is a mapping from $\mathbb{F}_{2^m}$ to itself satisfying the following necessary and sufficient conditions

$$F : z \to G(z) + \mu z \text{ is a permutation on } \mathbb{F}_{2^m}, \tag{2.8}$$

$$z \to F(z) + \beta z \text{ is 2-to-1 on } \mathbb{F}_{2^m} \text{ for any } \beta \in \mathbb{F}_{2^m}^*. \tag{2.9}$$

As proved in [37], condition (2.9) implies condition (2.8) and, thus, is necessary and sufficient for g being bent. Adding the linear term $\mathrm{tr}_m((\mu + 1)y)$ to (2.7) we obtain the original Dillon function (2.6). Therefore, functions in $\mathcal{H}$ and in the Dillon class are the same up to the addition of a linear term. It is observed in [37] that the class $\mathcal{H}$ contains all Niho type bent functions.

In [54] Dillon showed that the class H intersects with Maiorana-McFarland class and it has remained an open question whether H is contained in completed MM class. This problem is solved in [30] (see Sect. 4.4) by showing that the Niho bent functions of the first case do not belong to the completed MM class. Hence, Dillon's class H of bent functions is not contained in the completed MM class [30].

A natural extension of the class of bent functions is the class of plateaued functions. A Boolean function f on n variables is called *plateaued* if $\lambda_f(a) \in \{0, \pm\lambda\}$ for any $a \in \mathbb{F}_2^n$. The value λ is called the *amplitude* of the plateaued function. Because of (2.3) the amplitude λ cannot be null and must be a power 2^r, $\frac{n}{2} \leq r \leq n$. Bent functions are plateaued and, according to Parseval's relation (2.3), a plateaued function is bent if and only if its Walsh transform never takes the value 0. An (n, m)-function F is called *plateaued* if for any non-zero $c \in \mathbb{F}_2^m$ the Boolean function $c \cdot F$ is plateaued.

2.2.2   The Case of Odd Characteristics

Given a function f mapping $\mathbb{F}_{p^n}$ to $\mathbb{F}_p$ with p odd, its *Walsh transform* is defined as

$$\lambda_f(b) = \sum_{x \in \mathbb{F}_{p^n}} \omega^{f(x) - \mathrm{tr}_n(bx)}, \quad b \in \mathbb{F}_{p^n},$$

where $\omega = e^{\frac{2\pi i}{p}}$ is the complex primitive p^{th} root of unity and elements of $\mathbb{F}_p$ are considered as integers modulo p.

According to [84], a function f from $\mathbb{F}_{p^n}$ to $\mathbb{F}_p$ is called a *p-ary bent function* (or *generalized bent function*) if all its Walsh coefficients satisfy $|\lambda_f(b)|^2 = p^n$. A bent function f is called *regular* (see [80, 84]) if for every $b \in \mathbb{F}_{p^n}$ the normalized Walsh coefficient $p^{-n/2}\lambda_f(b)$ is equal to a complex p^{th} root of unity, i.e., $p^{-n/2}\lambda_f(b) = \omega^{f^*(b)}$ for some function f^* mapping $\mathbb{F}_{p^n}$ into $\mathbb{F}_p$. A bent function f is called *weakly regular* if there exists a complex u having unit magnitude such that $up^{-n/2}\lambda_f(b) = \omega^{f^*(b)}$ for all $b \in \mathbb{F}_{p^n}$. For a weakly regular function f, function f^* is called the *dual* of f. Recently, weakly regular bent functions were shown to be useful for constructing certain combinatorial objects such as partial difference sets, strongly regular graphs and association schemes (see [42, 102, 105]). This justifies why the classes of (weakly) regular bent functions are of independent interest.

It was long believed that all p-ary bent functions are weakly regular. However, some counter examples were found recently. In particular, ternary function f mapping $\mathbb{F}_{3^6}$ to $\mathbb{F}_3$ and given by $f(x) = \text{tr}_6(\alpha^7 x^{98})$ where α is a primitive element of $\mathbb{F}_{3^6}$, is bent and not weakly regular bent. An interesting open problem is to find an infinite class of non-weakly regular bent functions in a univariate representation.

Known univariate polynomials representing infinite classes of p-ary bent functions are listed in the table below. Here ξ denotes a primitive element of $\mathbb{F}_{3^n}$, "r" and "wr" refer to regular and weakly regular bent functions respectively. The first seven families in the table are monomials of the form $\text{tr}_n(ax^d)$ while the last one is a binomial bent function in the form $\text{tr}_n(F(x))$. Also, for any $a \in \mathbb{F}_{p^n}$, define the Kloosterman sum $K(a) = \sum_{x \in \mathbb{F}_{p^n}} w^{\text{tr}_n(x+ax^{-1})}$, where w is a complex p-th primitive root of unity. These functions were constructed and analyzed in [45, 68–76]. There are also numerous cases of quadratic functions and binomial ternary bent functions (see [69]).

Maiorana-McFarland construction of bent Boolean functions can be extended to the case of generalized bent functions. Let π be a permutation of $\mathbb{F}_p^m$ and $\sigma : \mathbb{F}_p^m \to \mathbb{F}_p$. Then $f : \mathbb{F}_p^m \times \mathbb{F}_p^m \to \mathbb{F}_p$ with $f(x, y) := x \cdot \pi(y) + \sigma(y)$ is a bent function. Moreover, the bijectiveness of π is necessary and sufficient for f being bent. Such bent functions are regular and the dual function is equal to $f^*(x, y) = y \cdot \pi^{-1}(x) + \sigma(\pi^{-1}(x))$. The completed MM class of generalized bent functions gives by far the widest class of bent functions, compared to all the other primary constructions.

A criterion for a function to be a member of the completed MM class in the binary case is given in [54]. In the general case, the proof is similar.

Proposition 2 [54] *Let n be an even positive integer, p any prime and $f : \mathbb{F}_p^n \mapsto \mathbb{F}_p$ a bent function. If f belongs to the completed MM class then there exists an $n/2$-dimensional vector subspace V in $\mathbb{F}_p^n$ such that the second order derivatives*

$$D_a D_c f(x) = f(x + a + c) - f(x + a) - f(x + c) + f(x)$$

vanish for any $a, c \in V$.

Table 2.1 Generalized bent functions

n	d or $F(x)$	a	d°	Remarks
	2	$a \neq 0$	2	r, wr
$2k$	$p^k + 1$	$a + a^{p^k} \neq 0$	2	wr
	$p^j + 1$, $\frac{n}{\gcd(n,j)}$-odd	$a \neq 0$	2	r, wr
	$p^j + 1$	Some condition on a	2	r, wr
	$\frac{3^k+1}{2}$, $\gcd(k,n) = 1$, k-odd	$a \neq 0$	$k + 1$	r, wr
$2k$	$t(3^k - 1)$, $\gcd(t, 3^k+1) = 1$	$K(a^{p^k+1}) = 0$	n	ternary r
$2k$	$\frac{3^n-1}{4} + 3^k + 1$, k-odd	$\xi^{\frac{3^k+1}{4}}$	n	ternary wr
$4k$	$x^{p^{3k}+p^{2k}-p^k+1} + x^2$		$(p-1)k + 2$	wr

Relation between the generalized bent functions of Table 2.1 and completed class of Maiorana-McFarland functions is studied in [29] (see also Sect. 4.5). In the binary case, the completed MM class contains all quadratic bent functions. However, this does not hold in the generalized case. First, for p odd there exist quadratic bent functions over $\mathbb{F}_{p^n}$ when n is odd while Maiorana-McFarland bent functions are defined only for n even. For the case n even, there also exist examples of quadratic generalized bent functions not belonging to the completed MM class [29]. Moreover, almost all of the non-quadratic classes in Table 2.1 do not intersect with the completed MM class [29]. This implies that in general, the Maiorana-McFarland construction is less overall than in the binary case even for the case n even.

2.3 APN and AB Functions

As we mentioned before perfect nonlinear or bent (n,m)-functions, being optimal against differential and linear attack, exist only for $m \leq n/2$. When $n = m$ functions with optimal resistance to differential and linear cryptanalysis are, respectively, almost perfect nonlinear and almost bent functions.

For any (n,m)-function F with $m \geq n$ the inequality

$$\mathcal{NL}(F) \leq 2^{n-1} - \frac{1}{2}(3 \cdot 2^n - 2(2^n - 1)(2^{n-1} - 1)/(2^m - 1) - 2)^{1/2}$$

gives a better upper bound for nonlinearity than the universal bound [39, 104]. This bound can be achieved only if $n = m$ with n odd when it takes the form

$$\mathcal{NL}(F) \leq 2^{n-1} - 2^{\frac{n-1}{2}}.$$

Functions achieving this bound are called *almost bent* (AB) or *maximum nonlinear*. AB functions are optimal against linear cryptanalysis. When n is even functions with the nonlinearity $2^{n-1} - 2^{\frac{n}{2}}$ are known and it is conjectured that this value is the highest possible nonlinearity for the case n even.

An (n, n)-function F is AB if and only if one of the following conditions is satisfied:

(i) $\Lambda_F = \{0, \pm 2^{\frac{n+1}{2}}\}$ [39];
(ii) for every $a, b \in \mathbb{F}_2^n$ the system of equations

$$\begin{cases} x + y + z & = a \\ F(x) + F(y) + F(z) & = b \end{cases}$$

has $3 \cdot 2^n - 2$ solutions (x, y, z) if $b = F(a)$, and $2^n - 2$ solutions otherwise [106];

(iii) the function $\gamma_F : \mathbb{F}_2^{2n} \to \mathbb{F}_2$ defined by the equality

$$\gamma_F(a, b) = \begin{cases} 1 & \text{if } a \neq 0 \text{ and } \delta_F(a, b) \neq 0 \\ 0 & \text{otherwise} \end{cases}$$

is bent [38].

For a function $F : \mathbb{F}_2^n \to \mathbb{F}_2^n$ and any elements $a, b \in \mathbb{F}_2^n$ we denote by $\delta_F(a, b)$ the number of solutions of the equation $F(x + a) + F(x) = b$, that is,

$$\delta_F(a, b) = |\{x \in \mathbb{F}_2^n : F(x + a) + F(x) = b\}|,$$

and we call the set
$$\Delta_F = \{\delta_F(a, b) : a, b \in \mathbb{F}_2^n, a \neq 0\}$$

the *differential spectrum* of the function F.

For any (n, n)-function F its differential uniformity

$$\delta_F = \max_{a, b \in \mathbb{F}_2^n, a \neq 0} \delta_F(a, b)$$

is not less than 2. Recall that F is almost perfect nonlinear (APN) if $\delta_F = 2$. APN functions possess the best resistance to the differential attack. The differential cryptanalysis presented by Biham and Shamir [7] is based on the study of how differences in an input can affect the resultant difference at the output. The resistance of a function F, used as an S-box in the cipher, to the differential attack is high when the value δ_F is small.

There are a few necessary and sufficient conditions for APN functions. Statements (i–iii) below easily follow from the definition of APN functions. An (n, n)-function F is APN if and only if one of the following conditions holds:

(i) $\Delta_F = \{0, 2\}$;
(ii) for any $a \in \mathbb{F}_2^{n*}$ the set

$$H_a = \{F(x + a) + F(x) : x \in \mathbb{F}_2^n\}$$

contains 2^{n-1} elements, that is $|H_a| = 2^{n-1}$;

(iii)   for every $(a, b) \neq 0$ the system

$$\begin{cases} x + y & = a \\ F(x) + F(y) & = b \end{cases}$$

admits 0 or 2 solutions;

(iv)   the function $\gamma_F : \mathbb{F}_2^{2n} \to \mathbb{F}_2$ defined by the equality

$$\gamma_F(a, b) = \begin{cases} 1 & \text{if } a \neq 0 \text{ and } \delta_F(a, b) \neq 0 \\ 0 & \text{otherwise} \end{cases}$$

has the weight $2^{2n-1} - 2^{n-1}$ [38];

(v) F is not affine on any 2-dimensional affine subspace of $\mathbb{F}_2^n$ [79].

For any (n, n)-function F we have the following inequality

$$\sum_{a,b \in \mathbb{F}_2^n} \lambda_F(a, b)^4 \geq 3 \cdot 2^{4n} - 2^{3n+1}$$

and the equality occurs if and only if F is APN. It easily follows that every AB function is APN ([39], see also [35]).

We know that the bentness of a function implies its perfect nonlinearity and vice versa. It is not quite the case with AB and APN functions. Not every APN function is AB. However, every quadratic APN function is AB (see [38]). For the general case there are some sufficient conditions for APN functions to be AB. For n odd, an APN function F is AB if and only if one of the following conditions is fulfilled [32]:

(i)   all the values in Λ_F are divisible by $2^{\frac{n+1}{2}}$;
(ii)   for any $c \in \mathbb{F}_2^{n*}$ the function $c \cdot F$ is plateaued.

Interesting subfamilies of APN functions are crooked and generalized crooked functions. An (n, n)-function F is called *crooked* if the following three conditions hold [3]:

1) $F(x) + F(y) + F(z) + F(x + y + z) \neq 0$ for any three distinct elements x, y, z,
2) $F(0) = 0$,
3) $F(x) + F(y) + F(z) + F(x + a) + F(y + a) + F(z + a) \neq 0$ for any $a \neq 0$ and x, y, z arbitrary.

Crooked functions form a subclass of AB permutations taking 0 value at 0 [3]. Every quadratic AB permutation taking 0 value at 0 is crooked. Every crooked function gives rise to a distance regular rectagraph [3]. We say that a function F is *generalized crooked* if the set $\{ u \in \mathbb{F}_{2^n} : F(x) + F(x + v) = u \text{ has solutions} \}$ is an affine hyperplane for any $v \neq 0$ (see [64, 65]). Every generalized crooked function is crooked, but the converse is not true. For instance, every quadratic APN function is generalized crooked.

2.3.1 The Case of Power Functions

There are natural reasons that in the beginning the main attention in the study of APN and AB functions was payed to power functions. AB power functions correspond to binary cyclic codes with two zeros, whose duals are optimal, and to pairs of maximum-length sequences (called M-sequences) with preferred crosscorrelation, which are used for spread-spectrum communications [38].

Checking APN and AB properties of power functions is easier than in the case of arbitrary polynomials. If F is a power function, that is $F(x) = x^d$, then F is APN if and only if the derivative $D_1 F$ is a two-to-one mapping. Indeed, since for any $a \neq 0$

$$D_a F(x) = (x + a)^d + x^d = a^d D_1 F(x/a)$$

then $D_a F$ is a two-to-one mapping if and only if $D_1 F$ is two-to-one.

Besides, the function $F(x) = x^d$ is AB if and only if $\lambda_F(a, b) \in \{0, \pm 2^{\frac{n+1}{2}}\}$ for $a \in \mathbb{F}_2, b \in \mathbb{F}_2^{n*}$, since $\lambda_F(a, b) = \lambda_F(1, a^{-d}b)$ for $a \in \mathbb{F}_2^{n*}$. In case F is a permutation, F is AB if and only if $\lambda_F(a, 1) \in \{0, \pm 2^{\frac{n+1}{2}}\}$ for $a \in \mathbb{F}_2^n$, since $\lambda_F(a, b) = \lambda_F(ab^{-\frac{1}{d}}, 1)$.

There are also simple sufficient condition for functions to be EA-inequivalent to power functions. If for an (n, n)-function F there exists an element $c \in \mathbb{F}_{2^n}^*$ such that $d^\circ(\text{tr}_n(cF)) \neq d^\circ(F)$ and $d^\circ(\text{tr}_n(cF)) > 1$, then F is EA-inequivalent to power functions [24]. Besides, for n odd, if an APN function F satisfies $d^\circ(F) \neq \min d^\circ(F)$ then F is EA-inequivalent to power functions [24].

The exponent d, $0 \leq d < 2^n - 1$, of a power function $F(x) = x^d$ on $\mathbb{F}_{2^n}$ gives an *equivalence class* (d) of exponents

$$(d) = \begin{cases} \{2^i d, & 2^i / d & : 0 \leq i < n\} & \text{if } x^d \text{ is a permutation} \\ \{2^i d & : 0 \leq i < n\} & \text{otherwise} \end{cases},$$

i.e. (d) is a union of 2-cyclotomic cosets of d and $\frac{1}{d}$ modulo $2^n - 1$ if x^d is a permutation, otherwise (d) is the 2-cyclotomic coset of d modulo $2^n - 1$. If d and d' belong to the same equivalence class then we call the power functions x^d and $x^{d'}$ *cyclotomic* equivalent. Obviously, if power functions F and F' are cyclotomic equivalent then $\Delta_F = \Delta_{F'}$ and $\Lambda_F = \Lambda_{F'}$.

Table 2.2 (resp. Table 2.3) gives all known values of exponents d (up to cyclotomic equivalence) such that the power function x^d is APN (resp. AB) and Table 2.4 gives all known values of d that x^d is a permutation with the best known nonlinearity (that is, $2^{n-1} - 2^{\frac{n}{2}}$) on the field $\mathbb{F}_{2^n}$ with n even. It is proved by Dobbertin that power APN functions are permutations when n is odd and 3-to-1 over $\mathbb{F}_{2^n}^*$ when n is even. When n is even, the inverse function x^{2^n-2} is a differentially 4-uniform permutation [98], and is chosen as the basic S-box, with $n = 8$, in the AES, see [48]. In [58] Dobbertin conjectured that Tables 2.2 and 2.3 represent the complete list of possible APN and AB power functions. This conjecture is confirmed for APN functions with $n \leq 25$ and for AB functions with $n \leq 33$ (see [60, 89]).

Since algebraic degree is an invariant for EA-equivalence and, in general, the functions of Table 2.2 (as well as their inverses) have different algebraic degrees,

Table 2.2 Known APN power functions x^d on $\mathbb{F}_{2^n}$

Functions	Exponents d	Conditions	$d^\circ(x^d)$	Proven
Gold	$2^i + 1$	$\gcd(i,n) = 1$	2	[67, 98]
Kasami	$2^{2i} - 2^i + 1$	$\gcd(i,n) = 1$	$i + 1$	[81, 82]
Welch	$2^t + 3$	$n = 2t + 1$	3	[59]
Niho	$2^t + 2^{\frac{t}{2}} - 1, \quad t$ even	$n = 2t + 1$	$(t+2)/2$	[58]
	$2^t + 2^{\frac{3t+1}{2}} - 1, t$ odd		$t + 1$	
Inverse	$2^{2t} - 1$	$n = 2t + 1$	$n - 1$	[4, 98]
Dobbertin	$2^{4i} + 2^{3i} + 2^{2i} + 2^i - 1$	$n = 5i$	$i + 3$	[60]

Table 2.3 Known AB power functions x^d on $\mathbb{F}_{2^n}$, n odd

Functions	Exponents d	Conditions	Proven
Gold	$2^i + 1$	$\gcd(i,n) = 1$	[67, 98]
Kasami	$2^{2i} - 2^i + 1$	$\gcd(i,n) = 1$	[82]
Welch	$2^t + 3$	$n = 2t + 1$	[31, 32]
Niho	$2^t + 2^{\frac{t}{2}} - 1, \quad t$ even	$n = 2t + 1$	[78]
	$2^t + 2^{\frac{3t+1}{2}} - 1, t$ odd		

Table 2.4 Known power permutations x^d with the highest known nonlinearity on $\mathbb{F}_{2^n}$, $n = 2t$

Exponents d	Conditions	Proven
$2^i + 1$	$\gcd(i,n) = 2, t$ odd	[67]
$2^{2i} - 2^i + 1$	$\gcd(i,n) = 2, t$ odd	[82]
$2^{n-1} - 1$		[85]
$2^t + 2^{\frac{t+1}{2}} + 1$	t odd	[47]
$2^t + 2^{t-1} + 1$	t odd	[47]
$2^t + 2^{\frac{t}{2}} + 1$	$t \equiv 2 \bmod 4$	[57]
$\sum_{k=0}^{t} 2^{ik}$	$\gcd(i,n) = 1, t$ even	[57, 96]

then no question about EA-equivalence between different functions of Table 2.2 could rise. Unlike EA-equivalence, CCZ-equivalence does not preserve algebraic degrees of functions. Thus, the question about CCZ-inequivalence of functions of Table 2.2 needs to be answered. It is proved in [26] (see also Sect. 3.5) that two Gold functions x^{2^i+1} and x^{2^j+1} with $1 \le i, j < n/2, i \ne j$, are CCZ-inequivalent, and that the Gold functions are CCZ-inequivalent to any Kasami and to the Welch functions (except in particular cases). Besides, the inverse and Dobbertin APN functions are CCZ-inequivalent to each other and to all other known power APN mappings [26] (see also Sect. 3.5). For all the other cases the problem stays open.

2.3.2 The Case of Polynomials

Before the work [24] the only known constructions of APN and AB functions were EA-equivalent to power functions, and it was widely accepted as true that all APN functions are EA-equivalent to power functions. Besides, CCZ-equivalence was considered as conjunction of EA-equivalence and taking inverses of permutations. In [24], it is proven that CCZ-equivalence is more general, and classes of APN and AB functions which are EA-inequivalent to power functions are constructed by applying CCZ-equivalence to the Gold APN and AB mappings. It was also shown in [24] that for $n = 5$ the constructed AB functions are EA-inequivalent to any permutations and this disproved the conjecture from [38] about nonexistence of such AB functions. It is proven recently in [91] that these AB functions are EA-inequivalent to permutations for any $n \geq 5$. However, it is still a question whether all AB functions are CCZ-equivalent to permutations. Table 2.5 presents the classes of APN functions constructed in [24]. When n is odd they are also AB. It is an open problem to determine whether the Gold power APN functions is the only case from Table 2.2 which allows construction of functions CCZ-equivalent but EA-inequivalent to them (see [90, 91]).

The new APN and AB functions introduced in [24] are, by construction, CCZ-equivalent to Gold functions. Hence, the problem of knowing whether there exist APN functions which would be CCZ-inequivalent to power functions remained open after their introduction. The first examples of APN functions CCZ-inequivalent to power functions where found in [65]. The first examples of such AB functions are constructed in [24] where we also present the first infinite families of such APN and AB polynomials. These functions are quadratic binomials given in Table 2.6 representing the families of APN and AB functions known nowadays. All these 11 families of APN functions are obviously AB when n is odd. When n is even they seem to have the same Walsh spectrum as Gold functions as already proven for the families (1–2), (5) and (8–11) [8–10, 12]. As proven by Carlet in [36], the families of APN functions (3), (4) and (11) from Table 2.6 are particular cases of a general construction (see Sect. 4.3).

Classification of APN functions is complete for $n \leq 5$ [16]: for these values of n the only APN functions, up to CCZ-equivalence, are power APN functions, and up to EA-equivalence, are power APN functions and those APN functions constructed in [24]. For $n = 6$ classification is complete for quadratic APN functions: 13 quadratic APN functions are found in [14] and, as proven in [63], up to CCZ-equivalence, these are the only quadratic APN functions. The only known APN function CCZ-inequivalent to power functions and to quadratic functions was found in [16, 64] for $n = 6$. For $n = 7$ and $n = 8$, as shown in a recent work [110], there are, respectively, more than 470 and more than 1000 CCZ-inequivalent quadratic APN functions.

APN Permutations One of the most important problems related to cryptographic functions is existence of APN permutations on $\mathbb{F}_{2^{2k}}$. It was conjectured that the answer is negative. Some nonexistence results were proven in [79, 99]: if F is a permutation on $\mathbb{F}_{2^{2k}}$ then it is not APN when one of the following conditions holds:

Table 2.5 Some APN functions CCZ-equivalent to Gold functions and EA-inequivalent to power functions on $\mathbb{F}_{2^n}$ (constructed in [24])

Functions	Conditions	d°
$x^{2^i+1} + (x^{2^i} + x + \mathrm{tr}_n(1) + 1)\mathrm{tr}(x^{2^i+1} + x\,\mathrm{tr}_n(1))$	$n \geq 4$ $\gcd(i,n) = 1$	3
$[x + \mathrm{tr}_{n/3}(x^{2(2^i+1)} + x^{4(2^i+1)}) + \mathrm{tr}(x)\mathrm{tr}_{n/3}(x^{2^i+1} + x^{2^{2i}(2^i+1)})]^{2^i+1}$	$6\|n$ $\gcd(i,n) = 1$	4
$x^{2^i+1} + \mathrm{tr}_{n/m}(x^{2^i+1}) + x^{2^i}\mathrm{tr}_{n/m}(x) + x\,\mathrm{tr}_{n/m}(x)^{2^i}$ $+[\mathrm{tr}_{n/m}(x)^{2^i+1}+\mathrm{tr}_{n/m}(x^{2^i+1})+\mathrm{tr}_{n/m}(x)]^{\frac{1}{2^i+1}}(x^{2^i}+\mathrm{tr}_{n/m}(x)^{2^i}+1)$ $+[\mathrm{tr}_{n/m}(x)^{2^i+1} + \mathrm{tr}_{n/m}(x^{2^i+1}) + \mathrm{tr}_{n/m}(x)]^{\frac{2^i}{2^i+1}}(x + \mathrm{tr}_{n/m}(x))$	$m \neq n$ n odd $m\|n$ $\gcd(i,n) = 1$	$m+2$

Table 2.6 Known classes of quadratic APN polynomials CCZ-inequivalent to power functions on $\mathbb{F}_{2^n}$

N°	Functions	Conditions	References
1-2	$x^{2^s+1} + \alpha^{2^k-1}x^{2^{ik}+2^{mk+s}}$	$n = pk, \gcd(k,3) = \gcd(s,3k) = 1,$ $p \in \{3,4\}, i = sk \bmod p, m = p - i,$ $n \geq 12, \alpha$ primitive in $\mathbb{F}_{2^n}^*$	Corol. 16, Thm. 24, and [25]
3	$x^{2^{2i}+2^i} + bx^{q+1} + cx^{q(2^{2i}+2^i)}$	$q = 2^m, n = 2m, \gcd(i,m) = 1,$ $\gcd(2^i + 1, q + 1) \neq 1, cb^q + b \neq 0,$ $c \notin \{\lambda^{(2^i+1)(q-1)}, \lambda \in \mathbb{F}_{2^n}\}, c^{q+1} = 1$	Corol. 23 and [18]
4	$x(x^{2^i} + x^q + cx^{2^i q})$ $+x^{2^i}(c^q x^q + sx^{2^i q})+x^{(2^i+1)q}$	$q = 2^m, n = 2m, \gcd(i,m) = 1,$ $c \in \mathbb{F}_{2^n}, s \in \mathbb{F}_{2^n} \backslash \mathbb{F}_q,$ $X^{2^i+1} + cX^{2^i} + c^q X + 1$ is irreducible over $\mathbb{F}_{2^n}$	Corol. 24 and [18]
5	$x^3 + a^{-1}\mathrm{tr}_n(a^3 x^9)$	$a \neq 0$	Corol. 25 and [28]
6	$x^3 + a^{-1}\mathrm{tr}_n^3(a^3 x^9 + a^6 x^{18})$	$3\|n, a \neq 0$	Corol. 32 and [27]
7	$x^3 + a^{-1}\mathrm{tr}_n^3(a^6 x^{18} + a^{12} x^{36})$	$3\|n, a \neq 0$	Corol. 32 and [27]
8–10	$ux^{2^s+1} + u^{2^k}x^{2^{-k}+2^{k+s}} +$ $vx^{2^{-k}+1} + wu^{2^k+1}x^{2^s+2^{k+s}}$	$n = 3k, \gcd(k,3) = \gcd(s,3k) = 1,$ $v, w \in \mathbb{F}_{2^k}, vw \neq 1,$ $3\|(k + s), u$ primitive in $\mathbb{F}_{2^n}^*$	[11]
11	$\alpha x^{2^s+1} + \alpha^{2^k}x^{2^{k+s}+2^k} +$ $\beta x^{2^k+1} + \sum_{i=1}^{k-1} \gamma_i x^{2^{k+i}+2^i}$	$n = 2k, \gcd(s,k) = 1, s, k$ odd, $\beta \notin \mathbb{F}_{2^k}, \gamma_i \in \mathbb{F}_{2^k},$ α not a cube	[11, 13]

(i) k is even and $F \in \mathbb{F}_{2^4}[x]$ [79];
(ii) F is a polynomial with coefficients in $\mathbb{F}_{2^k}$ [79];
(iii) F is a power function;
(iv) F is quadratic [99].

The conjecture on nonexistence of APN permutations on $\mathbb{F}_{2^{2k}}$ is disproved in [15]. Applying CCZ-equivalence to the trinomial APN function over $\mathbb{F}_{2^6}$ found in [14], Dillon et al. constructed an APN permutation over $\mathbb{F}_{2^6}$ [15]. This result is very important for future applications. However, it seems quite difficult to generalize it to a family and to extend this result to $k \geq 4$.

2.4 Commutative Presemifields and Semifields

In this subsection we are going to present commutative semifields and their important connection to quadratic planar functions [44, 49]. A ring with left and right distributivity and with no zero divisor is called a *presemifield*. A presemifield with a multiplicative identity is called a *semifield*. Any finite presemifield can be represented by $\mathbb{S} = (\mathbb{F}_{p^n}, +, \star)$, where p is a prime, n is a positive integer, $(\mathbb{F}_{p^n}, +)$ is the additive group of $\mathbb{F}_{p^n}$ and $x \star y = \phi(x, y)$ with ϕ a function from $\mathbb{F}_{p^n}^2$ onto $\mathbb{F}_{p^n}$, see [44, 83]. The prime p is called the *characteristic* of $\mathbb{S}$. Any finite field is a semifield. A semifield which is not a field is called *proper*.

Investigation of commutative semifields was launched by Dickson in 1906, shortly after the classification of finite fields. He constructed the first family of proper commutative semifields then [50, 51].

Let $\mathbb{S}_1 = (\mathbb{F}_{p^n}, +, \circ)$ and $\mathbb{S}_2 = (\mathbb{F}_{p^n}, +, \star)$ be two presemifields. They are called *isotopic* if there exist three linear permutations L, M, N over $\mathbb{F}_{p^n}$ such that

$$L(x \circ y) = M(x) \star N(y),$$

for any $x, y \in \mathbb{F}_{p^n}$. The triple (M, N, L) is called the *isotopism* between $\mathbb{S}_1$ and $\mathbb{S}_2$. If $M = N$ then $\mathbb{S}_1$ and $\mathbb{S}_2$ are called *strongly isotopic*.

Let $\mathbb{S} = (\mathbb{F}_{p^n}, +, \star)$ be a finite semifield. The subsets

$$N_l(\mathbb{S}) = \{\alpha \in \mathbb{S} : (\alpha \star x) \star y = \alpha \star (x \star y) \text{ for all } x, y \in \mathbb{S}\},$$

$$N_m(\mathbb{S}) = \{\alpha \in \mathbb{S} : (x \star \alpha) \star y = x \star (\alpha \star y) \text{ for all } x, y \in \mathbb{S}\},$$

$$N_r(\mathbb{S}) = \{\alpha \in \mathbb{S} : (x \star y) \star \alpha = x \star (y \star \alpha) \text{ for all } x, y \in \mathbb{S}\},$$

are called the *left, middle* and *right nucleus* of $\mathbb{S}$, respectively, and the set $N(\mathbb{S}) = N_l(\mathbb{S}) \cap N_m(\mathbb{S}) \cap N_r(\mathbb{S})$ is called the *nucleus*. These sets are finite fields and if $\mathbb{S}$ is commutative then $N_l(\mathbb{S}) = N_r(\mathbb{S}) \subseteq N_m(\mathbb{S})$. The nuclei measure how far $\mathbb{S}$ is from being associative. *The orders of the respective nuclei are invariant under isotopism* [44].

Every commutative presemifield can be transformed into a commutative semifield. Indeed, let $\mathbb{S} = (\mathbb{F}_{p^n}, +, \star)$ be a commutative presemifield which does not contain an identity. To create a semifield from $\mathbb{S}$, choose any $a \in \mathbb{F}_{p^n}^*$ and define a new multiplication $\circ$ by

$$(x \star a) \circ (a \star y) = x \star y$$

for all $x, y \in \mathbb{F}_{p^n}$. Then $\mathbb{S}' = (\mathbb{F}_{p^n}, +, \circ)$ is a commutative semifield isotopic to $\mathbb{S}$ with identity $a \star a$. We say $\mathbb{S}'$ is *a commutative semifield corresponding to the commutative presemifield* $\mathbb{S}$. An isotopism between $\mathbb{S}$ and $\mathbb{S}'$ is a strong isotopism $(L_a(x), L_a(x), x)$ with a linear permutation $L_a(x) = a \star x$, see [44].

Every commutative presemifield defines a planar DO polynomial and vice versa [44]. Let F be a quadratic PN function over $\mathbb{F}_{p^n}$. Then $\mathbb{S} = (\mathbb{F}_{p^n}, +, \star)$, with

$$x \star y = F(x + y) - F(x) - F(y)$$

for any $x, y \in \mathbb{F}_{p^n}$, is a commutative presemifield. We denote by $\mathbb{S}_F = (\mathbb{F}_{p^n}, +, \circ)$ the commutative semifield corresponding to the commutative presemifield $\mathbb{S}$ with isotopism $(L_1(x), L_1(x), x)$ and we call $\mathbb{S}_F = (\mathbb{F}_{p^n}, +, \circ)$ the *commutative semifield defined by the quadratic PN function* F. Conversely, given a commutative presemifield $\mathbb{S} = (\mathbb{F}_{p^n}, +, \star)$ of odd order, the function given by

$$F(x) = \frac{1}{2}(x \star x)$$

is a planar DO polynomial [44].

We have the following known facts on connection between CCZ-equivalence, isotopisms and strong isotopisms:

- two planar DO polynomials F and F' are CCZ-equivalent if and only if the corresponding commutative semifields $\mathbb{S}_F$ and $\mathbb{S}_{F'}$ are strongly isotopic [22];
- two commutative presemifields of order p^n with n odd are isotopic if and only if they are strongly isotopic [44];
- any commutative presemifield of odd order can generate at most two CCZ-equivalence classes of planar DO polynomials [44];
- if $\mathbb{S}_1$ and $\mathbb{S}_2$ are isotopic commutative semifields of characteristic p with the order of the middle nuclei and nuclei p^m and p^k, respectively, then one of the following statements must hold
  (a) m/k is odd and $\mathbb{S}_1$ and $\mathbb{S}_2$ are strongly isotopic,
  (b) m/k is even and either $\mathbb{S}_1$ and $\mathbb{S}_2$ are strongly isotopic or the only isotopisms between $\mathbb{S}_1$ and $\mathbb{S}_2$ are of the form $(\alpha \star N, N, L)$ where α is a non-square element of $N_m(\mathbb{S}_1)$ [44] ;
- there exist two commutative semifields of order 3^6 which are isotopic but not strongly isotopic [111].

Thus, according to the last point above, in the case n even it is possible that isotopic commutative presemifields define CCZ-inequivalent quadratic PN functions. However, isotopisms define an equivalence relation only over quadratic PN functions, and it is an open question whether this can be extended to an equivalence relation over all functions (from $\mathbb{F}_{p^n}$ to $\mathbb{F}_{p^m}$ for any positive integers n, m, and any prime p) preserving differential properties.

2.4.1 Known Cases of Planar Functions and Commutative Semifields

Almost all known planar functions are DO polynomials. The only known non-quadratic PN functions are the power functions

$$x^{\frac{3^t+1}{2}}$$

over $\mathbb{F}_{3^n}$, where t is odd and $\gcd(t,n) = 1$ [45, 74]. Although commutative semifields have been intensively studied for more than a hundred years, there are only a few cases of commutative semifields of odd order known (see [22, 44]). Below we present provably non-isotopic infinite families of commutative semifields (and corresponding planar functions) defined for any prime p known so far:

(i) $$x^2$$
over $\mathbb{F}_{p^n}$ which corresponds to the finite field $\mathbb{F}_{p^n}$;

(ii) $$x^{p^t+1}$$
over $\mathbb{F}_{p^n}$, with $n/\gcd(t,n)$ odd, which correspond to Albert's commutative twisted fields [1, 49, 75];

(iii) the functions over $\mathbb{F}_{p^{2k}}$, which correspond to the Dickson semifields [51];

(iv) the functions over $\mathbb{F}_{p^{2k}}$

$$(ax)^{p^s+1} - (ax)^{p^k(p^s+1)} + \sum_{i=0}^{k-1} c_i x^{p^i(p^k+1)}, \tag{2.10}$$

$$bx^{p^s+1} + (bx^{p^s+1})^{p^k} + cx^{p^k+1} + \sum_{i=1}^{k-1} r_i x^{p^{k+i}+p^i}, \tag{2.11}$$

where $a, b \in \mathbb{F}_{p^{2k}}^*$, b is not a square, $c \in \mathbb{F}_{p^{2k}} \setminus \mathbb{F}_{p^k}$, $r_i \in \mathbb{F}_{p^k}$, $0 \le i < k$, $\sum_{i=0}^{k-1} c_i x^{p^i}$ is a permutation of $\mathbb{F}_{p^k}$ with coefficients in $\mathbb{F}_{p^k}$, $\gcd(k+s, 2k) = \gcd(k+s, k)$, and for (2.11) also $\gcd(p^s+1, p^k+1) \ne \gcd(p^s+1, (p^k+1)/2)$ (see [21, 23]);

(v)
$$x^{p^s+1} - a^{p^t-1} x^{p^t+p^{2t+s}}$$

over $\mathbb{F}_{p^{3t}}$, where a is primitive in $\mathbb{F}_{p^{3t}}$, $\gcd(3,t) = 1$, $t - s = 0 \bmod 3$, $3t/\gcd(s, 3t)$ is odd (see [112]).

The following infinite families of commutative semifields (and corresponding planar functions) defined for any prime p were constructed recently but it is not known whether they are non-isotopic to the previously known families (1)–(v) above.

(vi)

$$x^{p^s+1} - a^{p^t-1} x^{p^{3t}+p^{t+s}}$$

over $\mathbb{F}_{p^{4t}}$, where a is primitive in $\mathbb{F}_{p^{4t}}$, $p^s \equiv p^t \equiv 1 \bmod 4$, $2t/\gcd(s, 2t)$ is odd (see [5]);

(vii) the function over $\mathbb{F}_{p^{2m}}$ for $m = 2k + 1$, $a \in \mathbb{F}_{p^2}^*$, (see [6, 22, 92])

$$a^{1-p} x^2 + x^{2p^m} + a^{1-p} \sum_{i=0}^{k} (-1)^i x^{p^{2i}(p^2+1)} + \sum_{j=0}^{k-1} (-1)^{k+j} x^{p^{2j+1}(p^2+1)}$$

$$- \left(\sum_{i=0}^{k} (-1)^i x^{p^{2i}(p^2+1)} + a^{p-1} \sum_{j=0}^{k-1} (-1)^{k+j} x^{p^{2j+1}(p^2+1)} \right)^{p^m},$$

for $a \in \mathbb{F}_{p^2} \setminus \mathbb{F}_p$ and $a \in \mathbb{F}_{p^*}$ it gives CCZ-inequivalent functions but the corresponding semifields are isotopic [22].

There are also a few cases of commutative semifields defined for $p = 3$:

(viii)

$$x^{10} \pm x^6 - x^2$$

over $\mathbb{F}_{3^n}$, with n odd, corresponding to the Coulter-Matthews and Ding-Yuan semifields [45, 56];

(ix) the function over $\mathbb{F}_{3^{2k}}$, with k odd, corresponding to the Ganley semifield [66];

(x) the function over $\mathbb{F}_{3^{2k}}$ corresponding to the Cohen-Ganley semifield [43];

(xi) the function over $\mathbb{F}_{3^{10}}$ corresponding to the Penttila-Williams semifield [100];

(xii) the function over $\mathbb{F}_{3^8}$ corresponding to the Coulter-Henderson-Kosick semifield [46];

(xiii)

$$x^2 + x^{90}$$

over $\mathbb{F}_{3^5}$ (see [107]).

The polynomial representations of functions (iii), (ix)-(xi) can be found in [95]. Note that PN functions (2.11) of family (iv) and families (v) and (vi) were constructed by following patterns of some known families of APN functions over fields of even characteristic, see [11, 25].

Further we have the following results on classification of commutative presemifields:

- any semifield of order p^2 is a finite field [83];
- any semifield of order p^3 is either a finite field or Albert's commutative twisted field [94];
- all planar DO functions over $\mathbb{F}_{3^5}$ are classified in [108]: there are 7 CCZ-inequivalent planar DO functions;
- a commutative presemifield which is three dimensional over its middle nucleus is necessarily isotopic to Albert's commutative twisted field [94];
- Albert's commutative twisted fields have left and middle nuclei of order $p^{\gcd(t,n)}$ [2];

- Dickson semifields of order p^{2k} have middle nuclei of order p^k [52];
- for $a \in \mathbb{F}_{p^k}$ the commutative semifields corresponding to the functions (2.11) of the family (iv) have middle nuclei of order p^d where d is even and divisible by $\gcd(s, k)$ [19];
- a DO polynomial (2.1) is CCZ-inequivalent to the planar function x^2 if $a_{jj} = 0$ for all j [22];
- a DO polynomial (2.1) is CCZ-inequivalent to the planar function x^{p^t+1}, with $n/\gcd(t, n)$ odd, if $a_{kj} = 0$ for all k and $j = k \pm t \bmod n$ [22].

References

1. A. A. Albert. On nonassociative division algebras. *Trans. Amer. Math. Soc.* 72, pp. 296–309, 1952.
2. A. A. Albert. Generalized twisted fields. *Pacific J. Math.* 11, pp. 1–8, 1961.
3. T. Bending, D. Fon-Der-Flaass. Crooked functions, bent functions and distance-regular graphs. *Electron. J. Comb.*, 5 (R34), 14, 1998.
4. T. Beth and C. Ding. On almost perfect nonlinear permutations. *Advances in Cryptology-EUROCRYPT'93, Lecture Notes in Computer Science*, 765, Springer-Verlag, New York, pp. 65–76, 1993.
5. J. Bierbrauer. New semifields, PN and APN functions. *Designs, Codes and Cryptography*, v. 54, pp. 189–200, 2010.
6. J. Bierbrauer. Commutative semifields from projection mappings. *Designs, Codes and Cryptography*, 61(2), pp. 187–196, 2011.
7. E. Biham and A. Shamir. Differential Cryptanalysis of DES-like Cryptosystems. *Journal of Cryptology* 4, no. 1, pp. 3–72, 1991.
8. C. Bracken, Z. Zha. On the Fourier Spectra of the Infinite Families of Quadratic APN Functions. *Finite Fields and Their Applications* 18(3), pp. 537–546, 2012.
9. C. Bracken, E. Byrne, N. Markin, G. McGuire. Determining the Nonlinearity of a New Family of APN Functions. *Applied Algebra, Algebraic Algorithms and Error Correcting Codes, Lecture Notes in Computer Science*, Vol 4851, Springer-Verlag, pp. 72–79, 2007.
10. C. Bracken, E. Byrne, N. Markin, G. McGuire. On the Walsh Spectrum of a New APN Function. *Cryptography and Coding*, Lecture Notes in Computer Science, Vol 4887, Springer-Verlag, pp. 92–98, 2007.
11. C. Bracken, E. Byrne, N. Markin, G. McGuire. New families of quadratic almost perfect nonlinear trinomials and multinomials. Finite Fields and Their Applications **14**(3), pp. 703–714, 2008.
12. C. Bracken, E. Byrne, N. Markin, G. McGuire. On the Fourier spectrum of Binomial APN functions. *SIAM journal of Discrete Mathematics*, 23(2), pp. 596–608, 2009.
13. C. Bracken, E. Byrne, N. Markin, G. McGuire. A Few More Quadratic APN Functions. *Cryptography and Communications* 3(1), pp. 43–53, 2011.
14. K. A. Browning, J. F. Dillon, R. E. Kibler, M. T. McQuistan. APN Polynomials and Related Codes. *Journal of Combinatorics, Information and System Science*, Special Issue in honor of Prof. D.K Ray-Chaudhuri on the occasion of his 75th birthday, vol. 34, no. 1–4, pp. 135–159, 2009.
15. K. A. Browning, J. F. Dillon, M. T. McQuistan, A. J. Wolfe. An APN Permutation in Dimension Six. *Post-proceedings of the 9-th International Conference on Finite Fields and Their Applications Fq'09, Contemporary Math.*, AMS, v. 518, pp. 33–42, 2010.
16. M. Brinkman and G. Leander. On the classification of APN functions up to dimension five. *Proceedings of the International Workshop on Coding and Cryptography 2007* dedicated to the memory of Hans Dobbertin, pp. 39–48, Versailles, France, 2007.

17. L. Budaghyan. The Simplest Method for Constructing APN Polynomials EA-Inequivalent to Power Functions. *Proceedings of First International Workshop on Arithmetic of Finite Fields, WAIFI 2007 Lecture Notes in Computer Science,* 4547, pp. 177–188, 2007.

18. L. Budaghyan and C. Carlet. Classes of Quadratic APN Trinomials and Hexanomials and Related Structures. *IEEE Trans. Inform. Theory,* vol. 54, no. 5, pp. 2354–2357, May 2008.

19. L. Budaghyan and C. Carlet. On CCZ-equivalence and its use in secondary constructions of bent functions. *Preproceedings of International Workshop on Coding and Cryptography WCC 2009,* pp. 19–36, 2009.

20. L. Budaghyan and C. Carlet. CCZ-equivalence of single and multi output Boolean functions. *Post-proceedings of the 9-th International Conference on Finite Fields and Their Applications Fq'09,* Contemporary Math., AMS, v. 518, pp. 43–54, 2010.

21. L. Budaghyan and T. Helleseth. New perfect nonlinear multinomials over $F_{p^{2k}}$ for any odd prime p. *Proceedings of the International Conference on Sequences and Their Applications SETA 2008,* Lecture Notes in Computer Science 5203, pp. 403–414, Lexington, USA, Sep. 2008.

22. L. Budaghyan and T. Helleseth. On Isotopisms of Commutative Presemifields and CCZ-Equivalence of Functions. Special Issue on Cryptography of *International Journal of Foundations of Computer Science,* v. 22(6), pp. 1243–1258, 2011. Preprint at http://eprint.iacr.org/2010/507

23. L. Budaghyan and T. Helleseth. New commutative semifields defined by new PN multinomials. *Cryptography and Communications: Discrete Structures, Boolean Functions and Sequences,* v. 3(1), pp. 1–16, 2011.

24. L. Budaghyan, C. Carlet, A. Pott. New Classes of Almost Bent and Almost Perfect Nonlinear Functions. *IEEE Trans. Inform. Theory,* vol. 52, no. 3, pp. 1141–1152, March 2006.

25. L. Budaghyan, C. Carlet, G. Leander. Two classes of quadratic APN binomials inequivalent to power functions. *IEEE Trans. Inform. Theory,* 54(9), pp. 4218–4229, 2008.

26. L. Budaghyan, C. Carlet, G. Leander. On inequivalence between known power APN functions. *Proceedings of the International Workshop on Boolean Functions: Cryptography and Applications, BFCA 2008,* Copenhagen, Denmark, May 2008.

27. L. Budaghyan, C. Carlet, G. Leander. On a construction of quadratic APN functions. *Proceedings of IEEE Information Theory Workshop, ITW'09,* pp. 374–378, Taormina, Sicily, Oct. 2009.

28. L. Budaghyan, C. Carlet, G. Leander. Constructing new APN functions from known ones. *Finite Fields and Their Applications,* v. 15, issue 2, pp. 150–159, April 2009.

29. L. Budaghyan, C. Carlet, T. Helleseth, A. Kholosha. Generalized Bent Functions and Their Relation to Maiorana-McFarland Class. *Proceedings of the IEEE International Symposium on Information Theory, ISIT 2012,* Cambridge, MA, USA, 1–6 July 2012.

30. L. Budaghyan, C. Carlet, T. Helleseth, A. Kholosha, S. Mesnager. Further Results on Niho Bent Functions. *IEEE Trans. Inform. Theory,* 58(11), pp. 6979–6985, 2012.

31. A. Canteaut, P. Charpin, H. Dobbertin. Weight divisibility of cyclic codes, highly nonlinear functions on $\mathbb{F}_{2^m}$, and crosscorrelation of maximum-length sequences. *SIAM Journal on Discrete Mathematics,* 13(1), pp. 105–138, 2000.

32. A. Canteaut, P. Charpin and H. Dobbertin. Binary m-sequences with three-valued cross-correlation: A proof of Welch's conjecture. *IEEE Trans. Inform. Theory,* 46 (1), pp. 4–8, 2000.

33. A. Canteaut, P. Charpin, and G. M. Kyureghyan, "A new class of monomial bent functions," *Finite Fields Appl.,* vol. 14, no. 1, pp. 221–241, Jan. 2008.

34. C. Carlet. Boolean Functions for Cryptography and Error Correcting Codes. Chapter of the monography *Boolean Methods and Models,* Yves Crama and Peter Hammer eds, Cambridge University Press, pp. 257–397, 2010.

35. C. Carlet. Vectorial Boolean Functions for Cryptography. Chapter of the monography *Boolean Methods and Models,* Yves Crama and Peter Hammer eds, Cambridge University Press, pp. 398–469, 2010.

36. C. Carlet. Relating three nonlinearity parameters of vectorial functions and building APN functions from bent functions. *Designs, Codes and Cryptography*, v. 59(1–3), pp. 89–109, 2011.

37. C. Carlet and S. Mesnager, "On Dillon's class H of bent functions, Niho bent functions and o-polynomials," *J. Combin. Theory Ser. A*, vol. 118, no. 8, pp. 2392–2410, Nov. 2011.

38. C. Carlet, P. Charpin and V. Zinoviev. Codes, bent functions and permutations suitable for DES-like cryptosystems. *Designs, Codes and Cryptography*, 15(2), pp. 125–156, 1998.

39. F. Chabaud and S. Vaudenay. Links between differential and linear cryptanalysis, *Advances in Cryptology -EUROCRYPT'94, LNCS*, Springer-Verlag, New York, 950, pp. 356–365, 1995.

40. P. Charpin, G. Kyureghyan. On a class of permutation polynomials over $\mathbb{F}_{2^n}$. *Proceedings of SETA 2008, Lecture Notes in Computer Science* 5203, pp. 368–376, 2008.

41. P. Charpin and G. M. Kyureghyan. Cubic monomial bent functions: A subclass of $\mathcal{M}$. *SIAM Journal on Discrete Mathematics*, vol. 22, no. 2, pp. 650–665, 2008.

42. Y. M. Chee, Y. Tan, and X. D. Zhang, "Strongly regular graphs constructed from p-ary bent functions," *J. Algebraic Combin.*, vol. 34, no. 2, pp. 251–266, Sep. 2011.

43. S. D. Cohen and M. J. Ganley. Commutative semifields, two-dimensional over there middle nuclei. *J. Algebra* 75, pp. 373–385, 1982.

44. R. S. Coulter and M. Henderson. Commutative presemifields and semifields. *Advances in Math.* 217, pp. 282–304, 2008.

45. R. S. Coulter and R. W. Matthews. Planar functions and planes of Lenz-Barlotti class II. *Des., Codes, Cryptogr.* 10, pp. 167–184, 1997.

46. R. S. Coulter, M. Henderson, P. Kosick. Planar polynomials for commutative semifields with specified nuclei. *Des. Codes Cryptogr.* 44, pp. 275–286, 2007.

47. T. Cusick and H. Dobbertin. Some new 3-valued crosscorrelation functions of binary m-sequences. *IEEE Trans. Inform. Theory*, 42, pp.1238–1240, 1996.

48. J. Daemen and V. Rijmen. AES proposal: Rijndael. http://csrc.nist.gov/encryption/aes/rijndael/Rijndael.pdf, 1999.

49. P. Dembowski and T. Ostrom. Planes of order n with collineation groups of order n^2. *Math. Z.* 103, pp. 239–258, 1968.

50. L. E. Dickson. Linear algebras in which division is always uniquely possible. *Trans. Amer. Math. Soc* 7, pp. 370–390, 1906.

51. L. E. Dickson. On commutative linear algebras in which division is always uniquely possible. *Trans. Amer. Math. Soc* 7, pp. 514–522, 1906.

52. L. E. Dickson. Linear algebras with associativity not assumed. *Duke Math. J.* 1, pp. 113–125, 1935.

53. J. F. Dillon. A survey of bent functions. *NSA Technical Journal Special Issue*, pp. 191–215, 1972.

54. J. F. Dillon. Elementary Hadamard Difference sets. Ph. D. Thesis, Univ. of Maryland, 1974.

55. J. F. Dillon and H. Dobbertin, "New cyclic difference sets with Singer parameters," *Finite Fields Appl.*, vol. 10, no. 3, pp. 342–389, Jul. 2004.

56. C. Ding and J. Yuan. A new family of skew Paley-Hadamard difference sets. *J. Comb. Theory Ser. A* 133, pp. 1526–1535, 2006.

57. H. Dobbertin. One-to-One Highly Nonlinear Power Functions on $GF(2^n)$. *Appl. Algebra Eng. Commun. Comput.* 9 (2), pp. 139–152, 1998.

58. H. Dobbertin. Almost perfect nonlinear power functions over $GF(2^n)$: the Niho case. *Inform. and Comput.*, 151, pp. 57–72, 1999.

59. H. Dobbertin. Almost perfect nonlinear power functions over $GF(2^n)$: the Welch case. *IEEE Trans. Inform. Theory*, 45, pp. 1271–1275, 1999.

60. H. Dobbertin. Almost perfect nonlinear power functions over $GF(2^n)$: a new case for n divisible by 5. *Proceedings of Finite Fields and Applications FQ5*, pp. 113–121, 2000.

61. H. Dobbertin. Private communication. 2004.

62. H. Dobbertin, G. Leander, A. Canteaut, C. Carlet, P. Felke, and P. Gaborit, "Construction of bent functions via Niho power functions," *J. Combin. Theory Ser. A*, vol. 113, no. 5, pp. 779–798, Jul. 2006.

63. Y. Edel. Quadratic APN functions as subspaces of alternating bilinear forms. *Contact Forum Coding Theory and Cryptography* III, Belgium (2009), pp. 11–24, 2011.

64. Y. Edel and A. Pott. A new almost perfect nonlinear function which is not quadratic. *Advances in Mathematics of Communications* 3, no. 1, pp. 59–81, 2009.

65. Y. Edel, G. Kyureghyan and A. Pott. A new APN function which is not equivalent to a power mapping. *IEEE Trans. Inform. Theory*, vol. 52, no. 2, pp. 744–747, Feb. 2006.

66. M. J. Ganley. Central weak nucleus semifields. *European J. Combin.* 2, pp. 339–347, 1981.

67. R. Gold. Maximal recursive sequences with 3-valued recursive crosscorrelation functions. *IEEE Trans. Inform. Theory*, 14, pp. 154–156, 1968.

68. G. Gong, T. Helleseth, H. Hu, and A. Kholosha. On the dual of certain ternary weakly regular bent functions. *IEEE Trans. Inf. Theory*, 58(4), pp. 2237–2243, 2012.

69. T. Helleseth and A. Kholosha. Monomial and quadratic bent functions over the finite fields of odd characteristic. *IEEE Trans. Inf. Theory*, vol. 52, no. 5, pp. 2018–2032, May 2006.

70. T. Helleseth and A. Kholosha. On the dual of monomial quadratic p-ary bent functions. *Sequences, Subsequences, and Consequences*, ser. Lecture Notes in Computer Science, S. Golomb, G. Gong, T. Helleseth, and H.-Y. Song, Eds., vol. 4893. Berlin: Springer-Verlag, 2007, pp. 50–61.

71. T. Helleseth and A. Kholosha. Sequences, bent functions and Jacobsthal sums. *Sequences and Their Applications—SETA 2010*, ser. Lecture Notes in Computer Science, C. Carlet and A. Pott, Eds., vol. 6338. Berlin: Springer-Verlag, 2010, pp. 416–429.

72. T. Helleseth and A. Kholosha. New binomial bent functions over the finite fields of odd characteristic. *IEEE Trans. Inf. Theory*, vol. 56, no. 9, pp. 4646–4652, Sep. 2010.

73. T. Helleseth and A. Kholosha. Crosscorrelation of m-sequences, exponential sums, bent functions and Jacobsthal sums. *Cryptography and Communications*, vol. 3, no. 4, pp. 281–291, Dec. 2011.

74. T. Helleseth and D. Sandberg. Some power mappings with low differential uniformity. *Applic. Alg. Eng., Commun. Comput.* 8, pp. 363–370, 1997.

75. T. Helleseth, C. Rong and D. Sandberg. New families of almost perfect nonlinear power mappings. *IEEE Trans. in Inf. Theory* 45, pp. 475–485, 1999.

76. T. Helleseth, H. D. L. Hollmann, A. Kholosha, Z. Wang, and Q. Xiang, "Proofs of two conjectures on ternary weakly regular bent functions," *IEEE Trans. Inf. Theory*, vol. 55, no. 11, pp. 5272–5283, Nov. 2009.

77. T. Helleseth, A. Kholosha, and S. Mesnager, "Niho bent functions and Subiaco hyperovals," in *Theory and Applications of Finite Fields*, ser. Contemporary Mathematics, M. Lavrauw, G. L. Mullen, S. Nikova, D. Panario, and L. Storme, Eds. Providence, Rhode Island: American Mathematical Society, 2012.

78. H. Hollmann and Q. Xiang. A proof of the Welch and Niho conjectures on crosscorrelations of binary m-sequences. *Finite Fields and Their Applications* 7, pp. 253–286, 2001.

79. X.-D. Hou. Affinity of permutations of $\mathbb{F}_2^n$. *Proceedings of the Workshop on the Coding and Cryptography 2003*, Augot, Charpin and Kabatianski eds, pp. 273–280, 2003.

80. X.-D. Hou.'p-Ary and q-ary versions of certain results about bent functions and resilient functions. *Finite Fields Appl.*, vol. 10, no. 4, pp. 566–582, Oct. 2004.

81. H. Janwa and R. Wilson. Hyperplane sections of Fermat varieties in P^3 in char. 2 and some applications to cyclic codes. *Proceedings of AAECC-10, LNCS*, vol. 673, Berlin, Springer-Verlag, pp. 180–194, 1993.

82. T. Kasami. The weight enumerators for several classes of subcodes of the second order binary Reed-Muller codes. *Inform. and Control*, 18, pp. 369–394, 1971.

83. D. E. Knuth. Finite semifields and projective planes. *J. Algebra* 2, pp. 182–217, 1965.

84. P. V. Kumar, R. A. Scholtz, and L. R. Welch, "Generalized bent functions and their properties," *J. Combin. Theory Ser. A*, vol. 40, no. 1, pp. 90–107, Sep. 1985.

85. G. Lachaud and J. Wolfmann. The Weights of the Orthogonals of the Extended Quadratic Binary Goppa Codes. *IEEE Trans. Inform. Theory*, vol. 36, pp. 686–692, 1990.

86. P. Langevin, G. Leander. Counting all bent functions in dimension eight 99270589265934370305785861242880. *Des. Codes Cryptography*59(1–3), pp. 193–205, 2011.

87. G. Leander. Monomial bent functions. *IEEE Transactions on Information Theory*, vol. 52, no. 2, pp. 738–743, 2006.
88. G. Leander and A. Kholosha, "Bent functions with 2^r Niho exponents," *IEEE Trans. Inf. Theory*, vol. 52, no. 12, pp. 5529–5532, Dec. 2006.
89. G. Leander and P. Langevin. On exponents with highly divisible Fourier coefficients and conjectures of Niho and Dobbertin. 2007.
90. Y. Li, M. Wang. Permutation polynomials EA-equivalent to the inverse function over $GF(2^n)$. *Cryptography and Communications* 3(3), pp. 175–186, 2011.
91. Y. Li, M. Wang. The Nonexistence of Permutations EA-Equivalent to Certain AB Functions. *IEEE Trans. Inf. Theory*, vol. 59, no. 1, pp. 672–679, 2013.
92. G. Lunardon, G. Marino, O. Polverion, R. Trombetti. Symplectic spreads and quadric Veroneseans. Manuscript, 2009.
93. M. Matsui. Linear cryptanalysis method for DES cipher. *Advances in Cryptology-EUROCRYPT'93, LNCS*, Springer-Verlag, pp. 386–397, 1994.
94. G. Menichetti. On a Kaplansky conjecture concerning three-dimensional division algebras over a finite field. *J. Algebra* 47, pp. 400–410, 1977.
95. K. Minami and N. Nakagawa. On planar functions of elementary abelian p-group type. Submitted.
96. Y. Niho. Multi-valued cross-correlation functions between two maximal linear recursive sequences. Ph.D. dissertation, Dept. Elec. Eng., Univ. Southern California. [*USCEE* Rep. 409], 1972.
97. K. Nyberg. Perfect nonlinear S-boxes. *Advances in Cryptography, EUROCRYPT'91*, Lecture Notes in Computer Science **547**, pp. 378–386, 1992.
98. K. Nyberg. Differentially uniform mappings for cryptography. *Advances in Cryptography, EUROCRYPT'93*, Lecture Notes in Computer Science 765, pp. 55–64, 1994.
99. K. Nyberg. S-boxes and Round Functions with Controllable Linearity and Differential Uniformity. *Proceedings of Fast Software Encryption 1994, LNCS* 1008, pp. 111–130, 1995.
100. T. Penttila and B. Williams. Ovoids of parabolic spaces. *Geom. Dedicata* 82, pp. 1–19, 2000.
101. A. Pott, Y. Zhou. CCZ and EA equivalence between mappings over finite Abelian groups. *Des. Codes Cryptography* 66(1–3), pp. 99–109, 2013.
102. A. Pott, Y. Tan, T. Feng, and S. Ling. Association schemes arising from bent functions. *Des. Codes Cryptogr.*, vol. 59, no. 1–3, pp. 319–331, Apr. 2011.
103. O. S. Rothaus. On "bent" functions. *J. Combin. Theory Ser. A*, vol. 20, no. 3, pp. 300–305, 1976.
104. V. Sidelnikov. On mutual correlation of sequences. *Soviet Math. Dokl.*, 12, pp. 197–201, 1971.
105. Y. Tan, A. Pott, and T. Feng. Strongly regular graphs associated with ternary bent functions. *J. Combin. Theory Ser. A*, vol. 117, no. 6, pp. 668–682, Aug. 2010.
106. E.R. van Dam, D. Fon-Der-Flaass. Codes, graphs, and schemes from nonlinear functions. *European J. Combin.* 24, 85–98, 2003.
107. G. Weng. Private communications, 2007.
108. G. Weng, X. Zeng. Further results on planar DO functions and commutative semifields. *Des. Codes Cryptogr.* 63, pp. 413–423, 2012.
109. S. Yoshiara. Equivalence of quadratic APN functions. *J. Algebr. Comb.* 35, pp. 461–475, 2012.
110. Y. Yu, M. Wang, Y. Li. A matrix approach for constructing quadratic APN functions. *Preproceedings of the International Conference WCC 2013,* Bergen, Norway, 2013.
111. Y. Zhou. A note on the isotopism of commutative semifields. Preprint, 2010.
112. Z. Zha, G. Kyureghyan, X. Wang. Perfect nonlinear binomials and their semifields. *Finite Fields and Their Applications* 15(2), pp. 125–133, 2009.

Chapter 3
Equivalence Relations of Functions

3.1 On the Structure of This Chapter

The notion of CCZ-equivalence, which seems the most natural among all equivalence notions in the block cipher framework and which seems to be also the most general, is difficult to handle, since checking whether two given functions are CCZ-equivalent or not is hard (at least when they share the same CCZ-invariant parameters). Building functions CCZ-equivalent (but not EA-equivalent) to a given function is hard too. The less general EA-equivalence is on the contrary simpler to check and, given some function, building EA-equivalent ones is very easy. Hence, identifying situations in which CCZ-equivalence reduces to EA-equivalence is useful. We show in this chapter that this happens for all single output Boolean functions and that it does not, for functions from $\mathbb{F}_p^n$ to $\mathbb{F}_p^m$ under condition that m is greater or equal to the smallest divisor of n different from 1, which for n even case simply implies $m \geq 2$ (see [2, 4]). In [18] these results were extended to a more general framework of functions over finite abelian groups in which the condition on m is reduced to $m \geq 2$ also for n odd case and for p odd also includes $m = 1$.

We prove further that CCZ-equivalence coincides with EA-equivalence for all (single output or multi ouput) PN (or bent) functions, and, more generally, for all functions whose all derivatives are surjective (see [1, 3, 5]).

Another question which has importance for theoretical and practical reasons is whether CCZ-equivalence is really the most general equivalence relation of functions which is relevant to the block cipher framework. We showed that trying to extend CCZ-equivalence to a more general notion in the same way as affine equivalence was extended to CCZ-equivalence (that is, by considering the CCZ-equivalence of the indicators of the graphs of the functions instead of that of the functions themselves) leads in fact to the same CCZ-equivalence (see [2, 4]).

Finally we study the question of CCZ-equivalence for known power APN functions and we prove that two Gold functions x^{2^i+1} and x^{2^j+1} with $1 \leq i, j < n/2$, $i \neq j$, are CCZ-inequivalent, and that the Gold functions are CCZ-inequivalent to any Kasami and to Welch functions (except in particular cases) [7]. We also show that the inverse and Dobbertin APN functions are CCZ-inequivalent to each other and to all other known power APN mappings [7].

© Springer International Publishing Switzerland 2014
L. Budaghyan, *Construction and Analysis of Cryptographic Functions*,
DOI 10.1007/978-3-319-12991-4_3

3.2   CCZ-Equivalence of (n, m)-Functions

We are going to investigate the question whether CCZ-equivalence of (n, m)-functions is strictly more general than their EA-equivalence. We already know that the answer to this question is yes when $n = m \geq 4$ since every permutation is CCZ-equivalent to its inverse [10], and, moreover, as shown in [6], when $n = m \geq 4$, CCZ-equivalence is still more general than the conjunction of EA-equivalence and of taking the inverse of a permutation. The question was open for general (n, m)-functions when $n \neq m$. In Sect. 3.2.1 we prove that the answer is also negative for (n, m)-functions when $m = 1$, that is, for Boolean functions. This poses then the question of knowing whether the case $m = 1$ is a particular case or if the same situation occurs for larger values of m. We give an almost complete answer to this question in Sect. 3.2.2 by showing that CCZ-equivalence of (n, m)-functions is strictly more general than their EA-equivalence when $n \geq 5$ and m is greater or equal to the smallest positive divisor of n different from 1.

3.2.1   The Case of Boolean Functions

Let us first consider the question whether CCZ-equivalence is strictly more general than EA-equivalence for Boolean functions. Let two Boolean functions f and f' of $\mathbb{F}_2^n$ be CCZ-equivalent but EA-inequivalent. Then, up to translation, there exist linear functions $L : \mathbb{F}_2^n \to \mathbb{F}_2^n$, and $l : \mathbb{F}_2^n \to \mathbb{F}_2$, and elements $a \in \mathbb{F}_2^n \setminus \{0\}$, $\eta \in \mathbb{F}_2$, such that

$$\mathcal{L}(x, y) = (L(x) + ay, l(x) + \eta y) \tag{3.1}$$

is a linear permutation of $\mathbb{F}_2^n \times \mathbb{F}_2$, and denoting:

$$F_1(x) = L(x) + af(x), \tag{3.2}$$

$$F_2(x) = l(x) + \eta f(x), \tag{3.3}$$

F_1 is a permutation of $\mathbb{F}_2^n$ and

$$f'(x) = F_2 \circ F_1^{-1}(x). \tag{3.4}$$

Hence we need characterizing the permutations of the form (3.2). Note that for any permutation (3.2) the linear function L must be either a permutation or 2-to-1. Thus, we have only two possibilities for the function F_1, that is, either

$$F_1(x) = L(x + L^{-1}(a)f(x))$$

when L is a permutation, or

$$F_1(x) = L'\big((x/b)^2 + x/b + L'^{-1}(a)f(x)\big) \tag{3.5}$$

when L is 2-to-1 and its kernel equals $\{0, b\}$, $b \in \mathbb{F}_{2^n}^*$, where L' is a linear permutation of $\mathbb{F}_{2^n}$ such that $L'((x/b)^2 + x/b) = L(x)$. Note that if we take $L^{-1} \circ F_1$ (L being a permutation) or $L'^{-1} \circ F_1$ (L being 2-to-1) in (3.4) instead of F_1, we get $f' \circ L$ and $f' \circ L'$, respectively, which are EA-equivalent to f'. Therefore, without loss of generality we can neglect L and L'. Then (3.5) gives (changing $L^{-1}(a)$ into a):

$$F_1(bx) = x^2 + x + ag(x) \tag{3.6}$$

where $g(x) = f(bx)$. Hence it is sufficient to consider permutations (3.2) of the following two types

$$x + af(x), \tag{3.7}$$

$$x^2 + x + af(x). \tag{3.8}$$

A lemma will simplify the study of these permutations:

Lemma 1 [2] *Let n be any positive integer, $a \in \mathbb{F}_{2^n}^*$ and f a Boolean function on $\mathbb{F}_{2^n}$.*
- The function

$$F(x) = x + af(x)$$

is a permutation over $\mathbb{F}_{2^n}$ if and only if F is an involution.
- The function

$$F'(x) = x + x^2 + af(x)$$

is a permutation over $\mathbb{F}_{2^n}$ if and only if $\mathrm{tr}_n(a) = 1$ and $f(x+1) = f(x)+1$ for every $x \in \mathbb{F}_{2^n}$. Under this condition, let H be any linear hyperplane of $\mathbb{F}_{2^n}$ not containing 1; for every $y \in \mathbb{F}_{2^n}$, there exists a unique element $\phi(y) \in \mathbb{F}_{2^n}$ such that

$$\phi(y) \in H \text{ and } \phi(y) + (\phi(y))^2 = y \quad \text{if } \mathrm{tr}_n(y) = 0,$$

$$\phi(y) = \phi(y+a) + 1 \quad\quad\quad\quad \text{if } \mathrm{tr}_n(y) = 1.$$

Then ϕ is a linear automorphism of $\mathbb{F}_{2^n}$ and we have

$$F'^{-1}(y) = \phi(y) + \mathrm{tr}_n(y) + f(\phi(y))$$

for every $y \in \mathbb{F}_{2^n}$.

Proof Let us assume that $F(x) = x + af(x)$ is a permutation. We have

$$F \circ F(x) = x + af(x) + af(x + af(x)).$$

If $f(x) = 0$ then obviously $F \circ F(x) = x$. If $f(x) = 1$ then $F \circ F(x) = x + a + af(x + a)$. Moreover, we have $f(x + a) = 1$ since otherwise $F(x + a) = F(x)$ which contradicts F being a permutation. Hence, when $f(x) = 1$, we have also $F \circ F(x) = x$. Therefore, $F^{-1} = F$.

If $F'(x) = x + x^2 + af(x)$ is a permutation over $\mathbb{F}_{2^n}$, then $\mathrm{tr}_n(a) = 1$ since otherwise we have $\mathrm{tr}_n(F'(x)) = 0$ for every $x \in \mathbb{F}_{2^n}$ (and F' is not surjective), and

$f(x + 1) = f(x) + 1$ for every x since if $f(x + 1) = f(x)$ for some $x \in \mathbb{F}_{2^n}$, then $F'(x + 1) = F'(x)$ and F' is not injective.

Conversely, if $\mathrm{tr}_n(a) = 1$ and $f(x + 1) = f(x) + 1$ for every $x \in \mathbb{F}_{2^n}$ then, for every $x, y \in \mathbb{F}_{2^n}$, we have $F'(x) = y$ if and only if:
- either $\mathrm{tr}_n(y) = f(x) = 0$ and x is the unique element outside $supp(f) = \{x \in \mathbb{F}_{2^n} / f(x) = 1\}$ such that $x + x^2 = y$;
- or $\mathrm{tr}_n(y) = f(x) = 1$ and x is the unique element of $supp(f)$ such that $x + x^2 = y + a$.

Hence, F' is a permutation over $\mathbb{F}_{2^n}$.

Moreover, assuming that this condition is satisfied, the relation $F'(x + 1) = F'(x) + a$, valid for every $x \in \mathbb{F}_{2^n}$, implies $F'^{-1}(y + a) = F'^{-1}(y) + 1$ for every $y \in \mathbb{F}_{2^n}$. The existence and uniqueness of $\phi(y)$ is straightforward. The restriction of ϕ to the hyperplane of equation $\mathrm{tr}_n(y) = 0$ is an isomorphism between this hyperplane and H. The restriction of ϕ to the hyperplane of equation $\mathrm{tr}_n(y) = 1$ is an isomorphism between this hyperplane and $\mathbb{F}_{2^n} \setminus H$. Hence ϕ is a linear automorphism of $\mathbb{F}_{2^n}$. Moreover, for every $x, y \in \mathbb{F}_{2^n}$, we have $F'(x) = y$ if and only if:
- either $\mathrm{tr}_n(y) = f(x) = 0$ and $x = \phi(y) + f(\phi(y))$ (indeed, if $\phi(y) \notin supp(f)$ then $\phi(y)$ is the unique element x of $\mathbb{F}_{2^n} \setminus supp(f)$ such that $x + x^2 = y$ and if $\phi(y) \in supp(f)$ then $\phi(y) + 1$ is the unique element x of $\mathbb{F}_{2^n} \setminus supp(f)$ such that $x + x^2 = y$ since $f(x + 1) = f(x) + 1$);
- or $\mathrm{tr}_n(y) = f(x) = 1$ and

$$x = F'^{-1}(y + a) + 1 = \phi(y + a) + f(\phi(y + a)) + 1 = \phi(y) + 1 + f(\phi(y)).$$

This completes the proof.   □

We deduce the main result of this subsection:

Theorem 1 [2] *Let n be any positive integer. Two Boolean functions of $\mathbb{F}_{2^n}$ are CCZ-equivalent if and only if they are EA-equivalent.*

Proof Let two Boolean functions f and f' on $\mathbb{F}_{2^n}$ be CCZ-equivalent and EA-inequivalent. Then there is a linear permutation $\mathcal{L}$ of $\mathbb{F}_{2^n}^2$ such that (3.1–3.4) take place. We first assume that $\eta = 1$.

If L is a permutation then, as mentioned above, without loss of generality we can assume $F_1(x) = x + af(x)$ and therefore $F_1^{-1} = F_1$ by Lemma 1. Then we get

$$f'(x) = l(F_1^{-1}(x)) + f(F_1^{-1}(x)) = l(x + af(x)) + f(x + af(x)).$$

If $f(x) = 0$ then $f'(x) = l(x)$. If $f(x) = 1$ then $f(x + a) = 1$ (see the proof of Lemma 1), and, therefore $f'(x) = l(x) + l(a) + 1$. Thus,

$$f'(x) = l(x) + (1 + l(a))f(x)$$

for every x. Note that $l(a) = 0$. Indeed, if $l(a) = 1$ then the system of equations

$$\begin{cases} x + ay = 0 \\ l(x) + y = 0 \end{cases}$$

has two solutions $(0, 0)$ and $(a, 1)$ which contradicts $\mathcal{L}$ being a permutation. Hence, $f'(x) = l(x) + f(x)$ and f is EA-equivalent to f', a contradiction.

Let now L be 2-to-1. Then, as observed above, we can assume without loss of generality that (3.6) takes place. Then, since $\mathcal{L}$ is bijective, we have $l(b) = 1$ (otherwise, the vector $(b, 0)$ would belong to the kernel of $\mathcal{L}$). By Lemma 1, we have $g(x + 1) = g(x) + 1$ for any $x \in \mathbb{F}_{2^n}$, that is, $f(bx + b) = f(bx) + 1$ for any $x \in \mathbb{F}_{2^n}$, that is, $f(x + b) = f(x) + 1$ for any $x \in \mathbb{F}_{2^n}$. By Lemma 1, the inverse of the function $x^2 + x + ag(x)$ equals $\phi(x) + \mathrm{tr}_n(x) + g(\phi(x))$ for a certain linear permutation ϕ of $\mathbb{F}_{2^n}$. Then, according to (3.6):

$$F_1^{-1}(x) = b(\phi(x) + \mathrm{tr}_n(x) + f(b\,\phi(x)))$$

and therefore, since $f' = F_2 \circ F_1^{-1}$:

$$f'(x) = l\big(b(\phi(x) + \mathrm{tr}_n(x) + f(b\,\phi(x)))\big) + f\big(b(\phi(x) + \mathrm{tr}_n(x) + f(b\,\phi(x)))\big)$$
$$= l(b\,\phi(x)) + \mathrm{tr}_n(x) + f(b\,\phi(x)) + f(b\,\phi(x)) + \mathrm{tr}_n(x) + f(b\,\phi(x))$$
$$= l(b\,\phi(x)) + f(b\,\phi(x)).$$

This means that f and f' are EA-equivalent, a contradiction.

We now assume that $\eta = 0$. According to the observations above and to Lemma 1, we can reduce ourselves to the cases $f'(x) = l(x + af(x))$ and $f'(x) = l\big(b(\phi(x) + \mathrm{tr}_n(x) + f(b\,\phi(x)))\big)$. For the first case we necessarily have $l(a) = 1$ and for the second case $l(b) = 1$ since otherwise the kernel of $\mathcal{L}$ would not be trivial (it would contain $(a, 1)$ and $(b, 0)$ respectively). Thus, $f'(x) = l(x) + f(x)$ or $f'(x) = l(b\,\phi(x)) + \mathrm{tr}_n(x) + f(b\,\phi(x))$, and therefore f and f' are EA-equivalent, a contradiction. $\square$

For any positive integers m and n, a Boolean function f of $\mathbb{F}_{2^n}$ can be considered as an (n, m)-function (since $\mathbb{F}_2$ is a subspace of $\mathbb{F}_{2^m}$). Hence it is a natural question whether an (n, m)-function f', which is CCZ-equivalent to f, is necessarily EA-equivalent to f. The theorem below shows that the answer is positive.

Theorem 2 [2] *Let m and n be any positive integers. Let f be a Boolean function of $\mathbb{F}_{2^n}$ and f' an (n, m)-function. Then f and f' are CCZ-equivalent as (n, m)-functions if and only if they are EA-equivalent as (n, m)-functions.*

Proof If f and f' are CCZ-equivalent as (n, m)-functions then there is a linear permutation $\mathcal{L}(x, y) = (L_1(x, y), L_2(x, y))$ of $\mathbb{F}_{2^n} \times \mathbb{F}_{2^m}$ such that $F_1(x) = L_1(x, f(x))$ is a permutation of $\mathbb{F}_{2^n}$ and $f' = F_2 \circ F_1^{-1}$ for $F_2(x) = L_2(x, f(x))$. As we saw above it is sufficient to consider only the cases

$$L_1(x, y) = x + ay, \tag{3.9}$$

$$L_1(x, y) = (x/b)^2 + x/b + ay, \tag{3.10}$$

where $a \in \mathbb{F}_{2^m}^*$, $b \in \mathbb{F}_{2^n}^*$. We have $L_2(x, y) = L'(x) + L''(y)$ for some linear functions $L' : \mathbb{F}_{2^n} \to \mathbb{F}_{2^m}$ and $L'' : \mathbb{F}_{2^m} \to \mathbb{F}_{2^m}$, and $F_2(x) = L'(x) + L''(f(x)) = L'(x) + L''(1)f(x)$. Since $\mathcal{L}$ is a permutation then the system

$$\begin{cases} x + ay = 0 \\ L'(x) + L''(y) = 0 \end{cases}$$

in case (3.9), and the system

$$\begin{cases} (x/b)^2 + x/b + ay = 0 \\ L'(x) + L''(y) = 0 \end{cases}$$

in case (3.10), must have only $(0,0)$ solution. Hence, $L'(a) \neq L''(1)$ for case (3.9) (since otherwise $(a, 1)$ is in the kernel of $\mathcal{L}$), and $L'(b) \neq 0$ for case (3.10) (since otherwise $(b, 0)$ is in the kernel of $\mathcal{L}$).

Using Lemma 1 in case (3.9) we get

$$\begin{aligned} f'(x) &= F_2 \circ F_1(x) \\ &= L'(x + af(x)) + L''(1)f(x + af(x)) \\ &= L'(x) + (L'(a) + L''(1))f(x) \end{aligned}$$

since $f(x + af(x)) = f(x)$ as we see it in the proof of Lemma 1. Hence f and f' are EA-equivalent as (n, m)-functions.

Applying Lemma 1 for case (3.10) we get

$$\begin{aligned} f'(x) &= F_2 \circ F_1^{-1}(x) \\ &= L'\big(b(\phi(x) + \mathrm{tr}_n(x) + f(b\,\phi(x)))\big) \\ &\quad + L''(1)f\big(b(\phi(x) + \mathrm{tr}_n(x) + f(b\,\phi(x)))\big) \\ &= L'(b\,\phi(x)) + L'(b)\mathrm{tr}_n(x) + L'(b)f(b\,\phi(x)) \\ &\quad + L''(1)f(b\,\phi(x)) + L''(1)\mathrm{tr}_n(x) + L''(1)f(b\,\phi(x)) \\ &= \big(L'(b\,\phi(x)) + L'(b)\mathrm{tr}_n(x) + L''(1)\mathrm{tr}_n(x)\big) + L'(b)f(b\,\phi(x)) \end{aligned}$$

since $f(x + b) = f(x) + 1$ as we see it from the proof of Lemma 1. Thus f and f' are EA-equivalent as (n, m)-functions.   □

Obviously, Theorem 2 is still valid if f is any (n, m)-function whose image set consists of only a pair of elements of $\mathbb{F}_2^m$, that is, if $|\{f(x) : x \in \mathbb{F}_2^n\}| = 2$.

Remark 1 [2] The paper [11] is dedicated to the study of permutations of the kind $G(x) + f(x)$ where f is a Boolean function of $\mathbb{F}_{2^n}$ and G is either a permutation or a linear function from $\mathbb{F}_{2^n}$ to itself. The results of this section, and Lemma 1 in particular, give a description of the inverses of all such permutations:

- Let L be a linear function from $\mathbb{F}_{2^n}$ to itself and f be a Boolean function of $\mathbb{F}_{2^n}$. If $F(x) = L(x) + f(x)$ is a permutation then F^{-1} is EA-equivalent to F.
- Let G be a permutation of $\mathbb{F}_{2^n}$ and f be a Boolean function of $\mathbb{F}_{2^n}$. If $F(x) = G(x) + f(x)$ is a permutation then $F^{-1}(x) = G^{-1}(x + f \circ G^{-1}(x))$.

The first assertion is straightforward and the second one is easily proved: we have $F(x) = H \circ G(x)$, where $H(x) = x + f \circ G^{-1}(x)$ is a permutation. H is involutive by Lemma 1; hence

$$F^{-1}(x) = G^{-1} \circ H^{-1}(x) = G^{-1} \circ H(x) = G^{-1}(x + f \circ G^{-1}(x)).$$

3.2.2   The Case of $m > 1$

We first show in Proposition 3 that for any divisor $m > 1$ of n, CCZ-equivalence of (n, m)-functions is strictly more general than EA-equivalence. Then, due to Proposition 4, we extend in Theorem 3 the hypotheses under which this is true.

Proposition 3 [2] *Let $n \geq 5$ and $m > 1$ be any divisor of n. Then for (n, m)-functions, CCZ-equivalence is strictly more general than EA-equivalence.*

Proof We need to treat the cases n odd and n even differently.

Let n be any odd positive integer, m any divisor of n and

$$F(x) = \operatorname{tr}_n^m(x^3). \tag{3.11}$$

The linear function $\mathcal{L}$ from $\mathbb{F}_{2^n} \times \mathbb{F}_{2^m}$ to itself

$$\mathcal{L}(x, y) = (L_1(x, y), L_2(x, y))$$
$$= \left(x + \operatorname{tr}_n(x) + \operatorname{tr}_m(y), y + \operatorname{tr}_n(x) + \operatorname{tr}_m(y)\right)$$

is an involution, and

$$F_1(x) = L_1(x, F(x)) = x + \operatorname{tr}_n(x) + \operatorname{tr}_n(x^3)$$

is an involution too (which is easy to check). Let

$$F_2(x) = L_2(x, F(x)) = \operatorname{tr}_n^m(x^3) + \operatorname{tr}_n(x) + \operatorname{tr}_n(x^3).$$

Then the function

$$F'(x) = F_2 \circ F_1^{-1}(x) = F_2 \circ F_1(x)$$
$$= \operatorname{tr}_n^m(x^3) + \operatorname{tr}_n^m(x^2 + x)\operatorname{tr}_n(x) + \operatorname{tr}_n^m(x^2 + x)\operatorname{tr}_n(x^3)$$

is CCZ-equivalent to F by definition. The part $\operatorname{tr}_{n/m}(x^2 + x)\operatorname{tr}_n(x^3)$ is nonquadratic for $n \geq 5$ and $m > 1$. Indeed, it equals

$$\sum_{\substack{0 \leq i < n \\ 0 \leq j < n/m}} x^{2^{i+1} + 2^i + 2^{jm}} + \sum_{\substack{0 \leq i < n \\ 0 \leq j < n/m}} x^{2^{i+1} + 2^i + 2^{jm+1}} \tag{3.12}$$

and for $n \geq 5$, $m > 1$, the item $x^{2^3 + 2^2 + 2^0}$ is cubic and does not vanish in (3.12). Hence, when $n \geq 5$ and $m > 1$ the (n, m)-functions F and F' have different algebraic degrees, and, therefore, they are EA-inequivalent while they are CCZ-equivalent by construction.

Let now n be any even positive integer, m any divisor of n and F be given by (3.11). The linear function

$$L(x, y) = (L_1(x, y), L_2(x, y)) = (x + \operatorname{tr}_m(y), y)$$

is an involution, and

$$F_1(x) = L_1(x, F(x)) = x + \mathrm{tr}_n(x^3)$$

is also involutive (this can be easily checked). Let

$$F_2(x) = L_2(x, F(x)) = \mathrm{tr}_n^m(x^3)$$

then

$$\begin{aligned}
F'(x) &= F_2 \circ F_1^{-1}(x) = F_2 \circ F_1(x) \\
&= \mathrm{tr}_n^m\big((x + \mathrm{tr}_n(x^3))^3\big) \\
&= \mathrm{tr}_n^m(x^3) + \mathrm{tr}_n^m(1)\mathrm{tr}_n(x^3) + \mathrm{tr}_n^m(x^2 + x)\mathrm{tr}_n(x^3).
\end{aligned}$$

The part $\mathrm{tr}_n^m(x^2+x)\,\mathrm{tr}_n(x^3)$ is nonquadratic when $n \geq 6$, $m > 1$, or when $n = m = 4$. Indeed, in these cases the item $x^{2^3+2^2+2^0}$ does not vanish in (3.12). Hence, the (n,m)-functions F and F' are CCZ-equivalent by construction, and when $n \geq 6$, $m > 1$, or when $n = m = 4$ they are EA-inequivalent because of the difference of their algebraic degrees. □

The next proposition will allow us to generalize the conditions under which the statement of Proposition 3 is valid.

Proposition 4 [2] *Let m and n be positive integers, and (n, m)-functions F and F' be CCZ-equivalent but EA-inequivalent. Then for any positive integer k and any (n,k)-function C there exists an (n,k)-function C' such that the $(n, m + k)$-functions $H(x) = (F(x), C(x))$ and $H'(x) = (F'(x), C'(x))$ are CCZ-equivalent and EA-inequivalent.*

Proof Let $L(x, y) = (L_1(x, y), L_2(x, y))$ be a linear permutation of $\mathbb{F}_{2^n} \times \mathbb{F}_{2^m}$ which maps the graph of F to the graph of F'. Then we have $F_1(x) = L_1(x, F(x))$, $F_2(x) = L_2(x, F(x))$, $F'(x) = F_2 \circ F_1^{-1}(x)$, where F_1 is a permutation. Let

$$\psi(x, (y, z)) = (\psi_1(x, (y, z)), \psi_2(x, (y, z)))$$

be the function from $\mathbb{F}_{2^n} \times \mathbb{F}_{2^m} \times \mathbb{F}_{2^k}$ to itself such that:

$$\psi_1(x, (y, z)) = L_1(x, y),$$
$$\psi_2(x, (y, z)) = (L_2(x, y), z).$$

The function ψ is linear and it is a permutation; indeed its kernel is the set of solutions of the system of two linear equations

$$\begin{cases} L_1(x, y) = 0 \\ (L_2(x, y), z) = (0, 0). \end{cases}$$

From the second equation we get $z = 0$ and we come down to the system

$$\begin{cases} L_1(x, y) = 0 \\ L_2(x, y) = 0. \end{cases}$$

which has the only solution $(0, 0)$. Hence the kernel of ψ is trivial.

For the $(n, m + k)$-function $H(x) = (F(x), C(x))$ denote $H_1(x) = \psi_1(x, H(x))$ and $H_2(x) = \psi_2(x, H(x))$ then

$$H_1(x) = \psi_1(x, H(x)) = \psi_1(x, (F(x), C(x))) = L_1(x, F(x)) = F_1(x)$$

which is a permutation and

$$H_2(x) = \psi_2(x, H(x)) = \psi_2(x, (F(x), C(x))) = (L_2(x, F(x)), C(x)) = (F_2(x), C(x)).$$

Hence, the function

$$H'(x) = H_2 \circ H_1^{-1}(x) = (F_2 \circ F_1^{-1}(x), C \circ F_1^{-1}(x)) = (F'(x), C'(x)),$$

where $C'(x) = C \circ F_1^{-1}(x)$, is CCZ-equivalent to $H(x)$. If F and F' are EA-inequivalent then obviously H and H' are EA-inequivalent too. $\square$

Obviously, Proposition 4 implies:

Corollary 1 [2] *Let n and m be any positive integers. If for (n, m)-functions CCZ-equivalence coincides with EA-equivalence then for (n, m')-functions, $1 \le m' \le m$, CCZ-equivalence coincides with EA-equivalence too.*

Proposition 3 and Proposition 4 give

Theorem 3 [2] *Let $n \ge 5$ and $k > 1$ be the smallest divisor of n. Then for any $m \ge k$ CCZ-equivalence of (n, m)-functions is strictly more general than EA-equivalence.*

In particular, Theorem 3 implies:

Corollary 2 [2] *If $n \ge 6$ is even then for every $m \ge 2$ CCZ-equivalence of (n, m)-functions is strictly more general than EA-equivalence.*

Remark 2 [2] CCZ-equivalence reduces to EA-equivalence for all (n, m)-functions with $n = 3$ and $1 \le m \le 3$ (the case $n = m = 3$ is checked with a computer and the rest follows from Corollary 1).

3.2.3   The Case of Bent Vectorial Boolean Functions

If two functions are CCZ-equivalent and one of them is bent then the second is bent too. Below we show that, in this framework, CCZ-equivalence coincides with EA-equivalence.

Theorem 4 [1] *Let n and m be positive integers and F be a bent function from $\mathbb{F}_2^n$ to $\mathbb{F}_2^m$. Then any function CCZ-equivalent to F is EA-equivalent to it.*

Proof Let F' be CCZ-equivalent to F and $\mathcal{L}(x, y) = (L_1(x, y), L_2(x, y))$, (with $L_1 : \mathbb{F}_2^n \times \mathbb{F}_2^m \to \mathbb{F}_2^n$, $L_2 : \mathbb{F}_2^n \times \mathbb{F}_2^m \to \mathbb{F}_2^m$) be an affine permutation of $\mathbb{F}_2^n \times \mathbb{F}_2^m$ which maps the graph of F to the graph of F'. Then $L_1(x, F(x))$ is a permutation and for some affine functions $L' : \mathbb{F}_2^n \to \mathbb{F}_2^n$ and $L'' : \mathbb{F}_2^m \to \mathbb{F}_2^n$ we can write $L_1(x, y) = L'(x) + L''(y)$.

For any element v of $\mathbb{F}_2^n$ we have

$$v \cdot L_1(x, F(x)) = v \cdot L'(x) + v \cdot L''(F(x)),$$

where "$\cdot$" is the inner product in $\mathbb{F}_2^n$ (if $\mathbb{F}_2^n$ is identified with $\mathbb{F}_{2^n}$, we can take $u \cdot v = \mathrm{tr}_n(uv)$ for any $u, v \in \mathbb{F}_{2^n}$). Since $L_1(x, F(x))$ is a permutation, then any function $v \cdot L_1(x, F(x))$ is balanced (recall that this property is a necessary and sufficient condition) and, hence, cannot be bent. Therefore, $v \cdot L''(F(x))$ cannot be bent either because $v \cdot L'(x)$ is an affine function. Then, the adjoint operator L''' of L'' (satisfying $v \cdot L''(F(x)) = L'''(v) \cdot F(x)$) is the null function since if $L'''(v) \neq 0$ then $L'''(v) \cdot F(x)$ is bent. This means that L'' is null, that is, L_1 depends only on x, which corresponds to EA-equivalence by Proposition 1.   □

Since the algebraic degree is preserved by EA-equivalence then Theorem 4 gives a very simple criterion for distinguishing inequivalent bent functions.

Corollary 3 [1] *Let n and m be any positive integers. If two bent (n, m)-functions have different algebraic degrees then they are CCZ-inequivalent.*

3.3   CCZ-Equivalence for Functions of Odd Characteristics

We are going to obtain an analogue of Theorem 3 for functions of odd characteristics.

Proposition 5 [4] *Let p be an odd prime, $n \geq 3$, and $m > 1$ be a divisor of n. Then there exist functions from $\mathbb{F}_{p^n}$ to $\mathbb{F}_{p^m}$ for which CCZ-equivalence is strictly more general than EA-equivalence.*

Proof The linear permutation of $\mathbb{F}_{p^n} \times \mathbb{F}_{p^m}$

$$\mathcal{L}(x, y) = (x + \mathrm{tr}_m(y), y)$$

maps the graph of a quadratic function $F : \mathbb{F}_{p^n} \to \mathbb{F}_{p^m}$

$$F(x) = \mathrm{tr}_n^m(x^2 - x^{p+1})$$

to the graph of a cubic function

$$F'(x) = \mathrm{tr}_n^m(x^2 - x^{p+1}) + \mathrm{tr}_n(x^2 - x^{p+1})\mathrm{tr}_n^m(x^p - x).$$

That is, the functions F and F' are CCZ-equivalent but EA-inequivalent.

Indeed, $\mathcal{L}$ is obviously a permutation since $(0,0)$ is the only solution of the system

$$x + \mathrm{tr}_m(y) = 0,$$
$$y = 0.$$

The function

$$F_1(x) = x + \mathrm{tr}_m(F(x)) = x + \mathrm{tr}_n(x^2 - x^{p+1})$$

is a permutation of $\mathbb{F}_{p^n}$ since for any $a \in \mathbb{F}_{p^n}^*$

$$F(x+a) - F(x) = x + a + \mathrm{tr}_n(x^2 + 2ax + a^2 - x^{p+1} - ax^p - a^p x - a^{p+1})$$
$$- x - \mathrm{tr}_n(x^2 - x^{p+1})$$
$$= a + \mathrm{tr}_n(a^2 - a^{p+1}) - \mathrm{tr}_n(x(a^p + a^{p^{n-1}} - 2a))$$

and the equality $F(x+a) = F(x)$ would imply $a + \mathrm{tr}_n(a^2 - a^{p+1}) = \mathrm{tr}_n(x(a^p + a^{p^{n-1}} - 2a))$, that is, $a \in \mathbb{F}_p^*$, that is, $a = 0$, a contradiction. Note further that the inverse of the function F_1 is

$$F_1^{-1}(x) = x - \mathrm{tr}_n(x^2 - x^{p+1})$$

since

$$F_1^{-1} \circ F_1(x) = x + \mathrm{tr}_n(x^2 - x^{p+1}) - \mathrm{tr}_n\left(x^2 + 2x\,\mathrm{tr}_n(x^2 - x^{p+1})\right.$$
$$+ \mathrm{tr}_n(x^2 - x^{p+1})^2 - x^{p+1} - x^p\,\mathrm{tr}_n(x^2 - x^{p+1})$$
$$\left. - x\,\mathrm{tr}_n(x^2 - x^{p+1})^p - \mathrm{tr}_n(x^2 - x^{p+1})^{p+1}\right)$$
$$= x - \mathrm{tr}_n(x^2 - x^{p+1})\,\mathrm{tr}_n(x - x^p) = x.$$

Hence, for $F_2(x) = F(x)$ we get

$$F_2 \circ F_1^{-1}(x) = \mathrm{tr}_n^m\left((x - \mathrm{tr}_n(x^2 - x^{p+1}))^2 - (x - \mathrm{tr}_n(x^2 - x^{p+1}))^{p+1}\right)$$
$$= \mathrm{tr}_n^m(x^2 - x^{p+1}) + \mathrm{tr}_n(x^2 - x^{p+1})\mathrm{tr}_n^m(x^p - x) = F'(x).$$

It is easy to check that for $m \geq 2$ and $n \geq 3$ the term x^{2p+1} has coefficient -2 in the polynomial representation of F'. Hence, F' has algebraic degree 3. By construction F and F' are CCZ-equivalent but they are EA-inequivalent because of the difference of their algebraic degrees. □

Next proposition is a restatement of Proposition 4 (given for binary case) for the case of any prime p. The proof for this general case is the same as in case $p = 2$ and we skip it.

Proposition 6 [4] *Let p be an odd prime, m and n positive integers, and functions F and F' from $\mathbb{F}_{p^n}$ to $\mathbb{F}_{p^m}$ be CCZ-equivalent but EA-inequivalent. Then for any positive integer k and any function C from $\mathbb{F}_{p^n}$ to $\mathbb{F}_{p^k}$ there exists a function C' from*

$\mathbb{F}_{p^n}$ to $\mathbb{F}_{p^k}$ such that the functions $H(x) = (F(x), C(x))$ and $H'(x) = (F'(x), C'(x))$ from $\mathbb{F}_{p^n}$ to $\mathbb{F}_{p^{m+k}}$ are CCZ-equivalent and EA-inequivalent.

Proposition 2 and Proposition 6 give

Theorem 5 [4] *Let p be an odd prime, $n \geq 3$ and $k > 1$ the smallest divisor of n. Then for any $m \geq k$, CCZ-equivalence of functions from $\mathbb{F}_{p^n}$ to $\mathbb{F}_{p^m}$ is strictly more general than their EA-equivalence.*

Remark 3 In [18] Theorems 3 and 5 are extended to a more general framework of finite abelian groups, in which the condition on m is relaxed: for the cases $p = 2$ it is sufficient to have $m \geq 2$ and for the rest of the cases $m \geq 1$.

3.3.1 CCZ-Equivalence and PN Functions

We exhibit below a large class of functions for which CCZ-equivalence coincides with EA-equivalence.

Theorem 6 [5, 3] *Let p be any prime, m and n any positive integers. If a function F from $\mathbb{F}_{p^n}$ to $\mathbb{F}_{p^m}$ is such that all its derivatives $D_a F(x) = F(x) - F(x + a), a \in \mathbb{F}_{p^n}^*$, are surjective, then any function CCZ-equivalent to F is EA-equivalent to it.*

Proof If functions F and F' are CCZ-equivalent then there exists an affine permutation $\mathcal{L}$ over $\mathbb{F}_{p^n} \times \mathbb{F}_{p^m}$ such that $\mathcal{L}(G_F) = G_{F'}$ where $G_F = \{(x, F(x)) \mid x \in \mathbb{F}_{p^n}\}$ and $G_{F'} = \{(x, F'(x)) \mid x \in \mathbb{F}_{p^n}\}$. The function $\mathcal{L}$ in this case can be introduced as

$$\mathcal{L}(x, y) = (L_1(x, y), L_2(x, y))$$

where $L_1 : \mathbb{F}_{p^n} \times \mathbb{F}_{p^m} \to \mathbb{F}_{p^n}$, $L_2 : \mathbb{F}_{p^n} \times \mathbb{F}_{p^m} \to \mathbb{F}_{p^m}$, are affine and $L_1(x, F(x))$ is a permutation. We are going to show that L_1 is independent of y when $D_a F(x) = F(x) - F(x + a), a \in \mathbb{F}_{p^n}^*$, are surjective. For some linear functions $L : \mathbb{F}_{p^n} \to \mathbb{F}_{p^n}$, $L' : \mathbb{F}_{p^m} \to \mathbb{F}_{p^n}$, and some $b \in \mathbb{F}_{p^n}$ we have

$$L_1(x, y) = L(x) + L'(y) + b.$$

If $L_1(x, F(x))$ is a permutation then for any $a \in \mathbb{F}_{p^n}^*$

$$L(x) + L'(F(x)) + b \neq L(x + a) + L'(F(x + a)) + b,$$

that is,

$$L'(F(x + a) - F(x)) \neq -L(a).$$

Since $F(x + a) - F(x)$ is surjective then the inequality above implies $L'(c) \neq L(a)$ for any $c \in \mathbb{F}_{p^m}$ and any $a \in \mathbb{F}_{p^n}^*$. First of all we see that L is a permutation, since otherwise $L(a') = 0 = L'(0)$ for some $a' \in \mathbb{F}_{p^n}^*$, and we get the inequality $L^{-1} \circ L'(c) \neq a$ for any $c \in \mathbb{F}_{p^m}$ and any $a \in \mathbb{F}_{p^n}^*$, which in its turn means $L^{-1} \circ L' = 0$, that is, $L' = 0$. Thus, F and F' are EA-equivalent by Proposition 1. □

Obviously, conditions of Theorem 6 are satisfied for PN functions:

Corollary 4 [5, 3] *Let F be a PN function and F' be CCZ-equivalent to F. Then F and F' are EA-equivalent.*

In particular, this corollary implies:

Corollary 5 [5, 3] *If a PN function F is CCZ-equivalent to a DO polynomial F' then F is also DO polynomial.*

Corollary 6 [5, 3] *Perfect nonlinear DO polynomials F and F' are CCZ-equivalent if and only if they are linear equivalent.*

3.4 Equivalence of Indicators of Graphs of Functions

It obviously follows from the definition of CCZ-equivalence that two functions are CCZ-equivalent if and only if the indicators of their graphs are affine equivalent. In the present section, we investigate whether CCZ-equivalence of the indicators of the graphs of functions can lead to a more general notion of equivalence of functions than CCZ-equivalence.

The Case of Even Characteristic For a given function F from $\mathbb{F}_2^n$ to $\mathbb{F}_2^m$ the indicator 1_{G_F} of its graph G_F is a Boolean function of $\mathbb{F}_2^{n+m}$. Hence, according to Theorem 1, for (n, m)-functions F and F' the indicators 1_{G_F} and $1_{G_{F'}}$ are CCZ-equivalent if and only if they are EA-equivalent. In the proposition below we prove that CCZ-equivalence of functions is the same as EA-equivalence of the indicators of the graphs of these functions.

Proposition 7 [2] *Let m and n be any positive integers. Two (n, m)-functions F and F' are CCZ-equivalent if and only if the indicators 1_{G_F} and $1_{G'_F}$ of their graphs are EA-equivalent.*

Proof It is obvious that when composing 1_{G_F} by an affine permutation $\mathcal{L}$ of $\mathbb{F}_2^{n+m}$ on the right, that is, taking $1_{G_F} \circ \mathcal{L}$, we are within the definition of CCZ-equivalence of functions, since $1_{G_F} \circ \mathcal{L} = 1_{\mathcal{L}^{-1}(G_F)}$. If we compose 1_{G_F} by an affine permutation $\mathcal{L}$ of $\mathbb{F}_2$ on the left, then we get $\mathcal{L} \circ 1_{G_F} = 1_{G_F} + b$ for $b \in \mathbb{F}_2$. Hence, we have only to prove that if for an (n, m)-function F' and for an affine Boolean function φ of $\mathbb{F}_2^{n+m}$, we have

$$1_{G_{F'}}(x, y) = 1_{G_F}(x, y) + \varphi(x, y)$$

then F and F' are CCZ-equivalent.

In case $m > 2$ we must have $\varphi = 0$ because 1_{G_F} and $1_{G_{F'}}$ have Hamming weight 2^n while, if φ is not null, it has then Hamming weight 2^{n+m-1} or 2^{n+m}, a contradiction, since $2^{n+m-1} > 2^{n+1}$. Thus, for $m > 2$ we get $F = F'$.

Let us consider now the case $m = 1$. Then $1_{G_F}(x, y) = F(x) + y + 1$ and $\varphi(x, y) = A(x) + ay + b$ for some affine Boolean function A of $\mathbb{F}_2^n$ and $a, b \in \mathbb{F}_2$.

Therefore,

$$1_{G_{F'}}(x, y) = 1_{G_F}(x, y) + \varphi(x, y) = F(x) + A(x) + (a + 1)y + b + 1.$$

If $a = 1$ then $1_{G_{F'}}$ is not the indicator of the graph of a function. Indeed, if $F(x_0) + A(x_0) = b$ for some $x_0 \in \mathbb{F}_2^n$ then $1_{G_{F'}}(x_0, 0) = 1_{G_{F'}}(x_0, 1) = 1$, a contradiction, and if such element x_0 does not exist then $F(x) + A(x) \equiv b + 1$ and $1_{G_{F'}}(x, y) \equiv 0$, a contradiction too.

If $a = 0$ then $1_{G_{F'}}(x, y) = 1$ if and only if $y = F(x) + A(x) + b$, that is, $F'(x) = F(x) + A(x) + b$ and F and F' are EA-equivalent and therefore CCZ-equivalent.

Let now $m = 2$. Then φ has Hamming weight 2^{n+1} while 1_{G_F} and $1_{G_{F'}}$ have Hamming weight 2^n. Therefore, for any $x \in \mathbb{F}_2^n$, we have $F(x) \neq F'(x)$ and

$$\varphi(x, y) = \begin{cases} 1 & \text{if } y \in \{F(x), F'(x)\} \\ 0 & \text{otherwise} \end{cases}.$$

Without loss of generality we can assume that $F(0) = 0$. Then $\varphi(0, 0) = \varphi(0, F(0)) = 1$ and $\varphi(0, F'(0)) = 1$. Since φ is affine then for any $x \in \mathbb{F}_2^n$

$$\varphi(x, F(x) + F'(0)) = \varphi(x, F(x)) + \varphi(0, F'(0)) + 1 = 1.$$

Thus, since $F'(0) \neq 0$, we get $F'(x) = F(x) + F'(0)$. $\square$

Due to Proposition 7 we can conclude:

Corollary 7 [2] *Let m and n be any positive integers. Two (n, m)-functions F and F' are CCZ-equivalent if and only if the indicators of their graphs 1_{G_F} and $1_{G_{F'}}$ are CCZ-equivalent.*

The Case of Odd Characteristics Below we prove that the equivalence of indicators of graphs of functions coincides with CCZ-equivalence for functions of odd characteristics as well. First we need some auxiliary results.

Lemma 2 [4] *Let p be an odd prime, n a positive integer, $a \in \mathbb{F}_{p^n}$ and f any function from $\mathbb{F}_{p^n}$ to itself with the image set $\{0, a\}$. If the function $F(x) = x + f(x)$ is a permutation of $\mathbb{F}_{p^n}$ then $x - f(x)$ is its inverse.*

Proof Denoting $F'(x) = x - f(x)$ we get

$$F' \circ F(x) = x + f(x) - f(x + f(x)).$$

If $f(x) = 0$ then obviously $F' \circ F(x) = x$.

If $f(x) = a$ then $F' \circ F(x) = x + a - f(x + a)$. Moreover, we have $f(x + a) = a$ since otherwise $F(x + a) = F(x)$ which contradicts F being a permutation. Hence, when $f(x) = a$, we have also $F' \circ F(x) = x$. Therefore, $F^{-1} = F'$. $\square$

As mentioned in [6], CCZ-equivalence can be considered not only for functions from $\mathbb{F}_{p^n}$ to itself but also for functions between arbitrary groups H_1 and H_2. In the following proposition we consider CCZ-equivalence of functions from $\mathbb{F}_{p^n}$ to $\mathbb{F}_2$.

Proposition 8 [4] *Let p be an odd prime and n a positive integer. Two functions f and f' from $\mathbb{F}_{p^n}$ to $\mathbb{F}_2$ are CCZ-equivalent if and only if $f' = f \circ A$ for some affine permutation A of $\mathbb{F}_{p^n}$.*

Proof Let the functions f and f' be CCZ-equivalent. Then there exists an affine permutation $\mathcal{L}$ of $\mathbb{F}_{p^n} \times \mathbb{F}_2$ such that $\mathcal{L}(G_f) = G_{f'}$. Without loss of generality we can assume that $\mathcal{L}$ is linear. Then there exist linear functions $L : \mathbb{F}_{p^n} \to \mathbb{F}_{p^n}$, $\phi : \mathbb{F}_2 \to \mathbb{F}_{p^n}$, $l : \mathbb{F}_{p^n} \to \mathbb{F}_2$ and an element $a \in \mathbb{F}_2$ such that

$$\mathcal{L}(x, y) = (L(x) + \phi(y), l(x) + ay),$$

and for

$$F_1(x) = L(x) + \phi \circ f(x),$$
$$F_2(x) = l(x) + af(x),$$

F_1 is a permutation of $\mathbb{F}_{p^n}$ and

$$f'(x) = F_2 \circ F_1^{-1}(x).$$

Note that any linear function l from $\mathbb{F}_p^n$ to $\mathbb{F}_2$ must be 0 since otherwise it is balanced, which is impossible since p^n is an odd number. Hence, we have $l(x) = 0$ and, since $\mathcal{L}$ is a permutation, $a = 1$, that is, $F_2(x) = f(x)$. Besides, if $\phi \circ f = 0$ then obviously L is a permutation and $f' = f \circ L^{-1}$ and we can take $A = L^{-1}$. Hence we assume that ϕ has the image set $\{0, b\}$ where $b \neq 0$ and $\phi \circ f$ is not a zero function.

Since F_1 is a permutation and the image of $\phi \circ f$ consists of 2 elements then the function L must have at most 2 zeros, and, since $p \geq 3$ and L is a linear function from $\mathbb{F}_{p^n}$ to itself then it has exactly one zero, that is, L is a permutation. Hence,

$$F_1(x) = L(x + L^{-1} \circ \phi \circ f(x)),$$

where the function $F_1^*(x) = x + L^{-1} \circ \phi \circ f(x)$ is a permutation too, and therefore, by Lemma 2 its inverse is $F_1^{*-1}(x) = x - L^{-1} \circ \phi \circ f(x)$. We get

$$F_1^{-1}(x) = F_1^{*-1} \circ L^{-1}(x)$$

and then

$$f' \circ L(x) = F_2 \circ F_1^{*-1}(x) = f(x - L^{-1} \circ \phi \circ f(x)).$$

If $f(x) = 0$ then $f' \circ L(x) = 0 = f(x)$.

If $f(x) = 1$ then we have $f(x - L^{-1}(b)) = 1 = f(x)$. Indeed, if $f(x) = 1$ and $f(x - L^{-1}(b)) = 0$ then

$$F^{*-1}(x - L^{-1}(b)) = x - L^{-1}(b) - L^{-1} \circ \phi \circ f(x - L^{-1}(b)) = x - L^{-1}(b),$$

$$F^{*-1}(x) = x - L^{-1} \circ \phi \circ f(x) = x - L^{-1}(b),$$

which contradict F^{*-1} being a permutation. Hence, $f' \circ L(x) = f(x)$ and we can take $A = L^{-1}$. $\qquad \square$

Now we can prove the main result of this section:

Theorem 7 [4] *Let n and m be any positive integers, p any prime, and F and F′ any functions from $\mathbb{F}_{p^n}$ to $\mathbb{F}_{p^m}$. Then F and F′ are CCZ-equivalent if and only if the indicators of their graphs 1_{G_F} and $1_{G_{F'}}$ are CCZ-equivalent.*

Proof For the case $p = 2$ this theorem states Corollary 7. Let p be odd. Since 1_{G_F} and $1_{G_{F'}}$ are functions from $\mathbb{F}_{p^n} \times \mathbb{F}_{p^n}$ to $\mathbb{F}_2$ then according to Proposition 8 they are CCZ-equivalent if and only if there exists an affine permutation A of $\mathbb{F}_{p^n} \times \mathbb{F}_{p^n}$ that $1_{G_{F'}} = 1_{G_F} \circ A$, that is, if and only if F and F' are CCZ-equivalent.   □

3.5 CCZ-Equivalence and Power APN Functions

It is clear that the known power APN functions are pairwise EA-inequivalent since in general these functions (as well as their inverses) have different algebraic degrees. Unlike EA-equivalence, CCZ-equivalence does not preserve algebraic degrees of functions. Thus, it is an open question whether the known power APN functions (functions of Table 2.2) are pairwise CCZ-inequivalent. Below we solve this problem for some cases. We prove that two Gold functions x^{2^i+1} and x^{2^j+1} with $1 \leq i, j < n/2, i \neq j$, are CCZ-inequivalent, and that the Gold functions are CCZ-inequivalent to any Kasami and to the Welch functions (except in particular cases). We also note that the inverse and Dobbertin APN functions are CCZ-inequivalent to all known power APN mappings because of their unique Walsh spectra.

Proving CCZ-inequivalence of two functions can be done, in some cases, by showing that some invariants (such as, for instance, the extended Walsh spectrum) are different for the two functions. However, we could not find such invariants in the general cases of inequivalence we study here (except for the inverse and Dobbertin functions). So we give brute-force proofs: supposing that the functions are equivalent, we show this leads to contradictions.

Without loss of generality a Gold function $F(x) = x^{2^s+1}$ and a Kasami function $K(x) = x^{4^r-2^r+1}$ can be considered under conditions $1 \leq s < \frac{n}{2}, 2 \leq r < \frac{n}{2}$, since this exhausts all different cases (under EA-equivalence).

CCZ-Inequivalence of Two Gold Functions

Theorem 8 [7] *Let $F(x) = x^{2^s+1}$, $G(x) = x^{2^r+1}$ and $s \neq r$, $1 \leq s, r < \frac{n}{2}$, $\gcd(s, n) = \gcd(r, n) = 1$. Then F and G are CCZ-inequivalent on $\mathbb{F}_{2^n}$.*

Proof Suppose that $F(x)$ and $G(x)$ are CCZ-equivalent, then there exists an affine automorphism $\mathcal{L} = (L_1, L_2)$ of $\mathbb{F}_{2^n} \times \mathbb{F}_{2^n}$ such that

$$L_2(x, F(x)) = G(L_1(x, F(x))). \tag{3.13}$$

Writing

$$L_1(x, y) = L(x) + L'(y), \tag{3.14}$$

$$L_2(x, y) = L''(x) + L'''(y), \tag{3.15}$$

$$L(x) = b + \sum_{m \in \mathbb{Z}/n\mathbb{Z}} b_m x^{2^m}, \tag{3.16}$$

$$L'(x) = b' + \sum_{m \in \mathbb{Z}/n\mathbb{Z}} b'_m x^{2^m}, \tag{3.17}$$

$$L''(x) = b'' + \sum_{m \in \mathbb{Z}/n\mathbb{Z}} b''_m x^{2^m}, \tag{3.18}$$

$$L'''(x) = b''' + \sum_{m \in \mathbb{Z}/n\mathbb{Z}} b'''_m x^{2^m}, \tag{3.19}$$

$$b + b' = c. \tag{3.20}$$

$$b'' + b''' = c', \tag{3.21}$$

we can rewrite (3.13) as

$$L''(x) + L'''(F(x)) = G[L(x) + L'(F(x))]. \tag{3.22}$$

Considering the right side of equality (3.22) we get

$$G[L(x) + L'(F(x))] = \left(L(x) + L'(x^{2^s+1}) \right) \left(L(x) + L'(x^{2^s+1}) \right)^{2^r}$$

$$= \left(c + \sum_{m \in \mathbb{Z}/n\mathbb{Z}} b_m x^{2^m} + \sum_{m \in \mathbb{Z}/n\mathbb{Z}} b'_m x^{2^m(2^s+1)} \right)$$

$$\times \left(c^{2^r} + \sum_{m \in \mathbb{Z}/n\mathbb{Z}} b_m^{2^r} x^{2^{m+r}} + \sum_{m \in \mathbb{Z}/n\mathbb{Z}} b'^{2^r}_m x^{2^{m+r}(2^s+1)} \right)$$

$$= \sum_{m,k \in \mathbb{Z}/n\mathbb{Z}} b_k b_m^{2^r} x^{2^{m+r}(2^s+1)+2^k} + \sum_{m,k \in \mathbb{Z}/n\mathbb{Z}} b'_k b_m^{2^r} x^{2^{m+r}+2^k(2^s+1)}$$

$$+ \sum_{m,k \in \mathbb{Z}/n\mathbb{Z}} b'_k b'^{2^r}_m x^{2^{m+r}(2^s+1)+2^k(2^s+1)} + Q(x),$$

where $Q(x)$ is a quadratic polynomial. Since the left side of equality (3.22) is a quadratic polynomial then all terms in the expression above whose exponents have 2-weight strictly greater than 2 must cancel.

One of the functions L and L' must be non-constant because otherwise $L_1(x, F(x))$ cannot be a permutation, and this contradicts the assumption about CCZ-equivalence of F and G. Let $L' \neq \text{const}$ then there exists $m \in \mathbb{Z}/n\mathbb{Z}$ such that $b'_m \neq 0$. Considering s and r as elements of $\mathbb{Z}/n\mathbb{Z}$ we have $s \neq \pm r$ and $s, r \neq 0$. Then

$2^{m+r}(2^s + 1) + 2^m(2^s + 1)$ has 2-weight 4 and the items with this exponent have to vanish. We get $b'^{2^r+1}_m + b'_{m+r}b'^{2^r}_{m-r} = 0$ and since $b'_m \neq 0$ then $b'_{m+r}, b'_{m-r} \neq 0$ and $b'_m b'^{-2^r}_{m-r} = b'_{m+r}b'^{-2^r}_m$. Since $\gcd(r, n) = 1$ then applying this observation for $m + r$, $m + 2r,...,$ instead of m we get $b'_t \neq 0$ and, for some $\lambda \in \mathbb{F}_{2^n}$,

$$\lambda = b'_m b'^{-2^r}_{m-r} = b'_{t+r}b'^{-2^r}_t \tag{3.23}$$

for all $t \in \mathbb{Z}/n\mathbb{Z}$.

Let us consider the sum

$$\sum_{m,k \in \mathbb{Z}/n\mathbb{Z}} b'_k b'^{2^r}_m x^{2^{m+r}(2^s+1)+2^k(2^s+1)}.$$

For any $k, m \in \mathbb{Z}/n\mathbb{Z}$, $k \neq m + r$, the items $b'_k b'^{2^r}_m x^{2^{m+r}(2^s+1)+2^k(2^s+1)}$ and $b'_{m+r}b'^{2^r}_{k-r}x^{2^k(2^s+1)+2^{m+r}(2^s+1)}$ differ and cancel pairwise because of (3.23). In the case $k = m + r$ the sum gives items with the exponents of 2-weight not greater than 2.

Equality (3.23) implies $b'_{t+r} = \lambda b'^{2^r}_t$ for all t. Then, introducing μ such that $\lambda = \mu^{2^r-1}$, we deduce that $\mu b'_{t+r} = (\mu b'_t)^{2^r}$ for all t and then that $\mu b'_{t+1} = (\mu b'_t)^2$ (using that $\gcd(r, n) = 1$) and then $\mu b'_t = (\mu b'_0)^{2^t}$. This means that

$$\mu L'(x) = \mu b' + \text{tr}(\mu b'_0 x). \tag{3.24}$$

Then obviously L' is not a permutation and, since $L_1(x, F(x))$ is a permutation, then L is not a constant. Thus $b_m \neq 0$ for some $m \in \mathbb{Z}/n\mathbb{Z}$.

Since $s \neq \pm r$ then considering the items with the exponent $2^{m+r+s} + 2^{m+r} + 2^m$ we get $b_m b'^{2^r}_m + b'_{m+r}b^{2^r}_{m-r} = 0$ if $r \neq -2s$. Since $b_m, b'_m \neq 0$ then $b_{m-r} \neq 0$ and $b_m b^{-2^r}_{m-r} = b'_{m+r}b'^{-2^r}_m$. Repeating these steps for $b_{m-r}, b_{m-2r}, ...,$ because of (3.23) we get $b_t \neq 0$ for all $t \in \mathbb{Z}/n\mathbb{Z}$ and

$$\lambda = b'_m b'^{-2^r}_{m-r} = b_t b^{-2^r}_{t-r}. \tag{3.25}$$

For the case $r = -2s$ (since $\gcd(s, n) = 1$) we can consider the items with the exponent $2^{m+r} + 2^{m+s} + 2^m$ and get $b'_m b^{2^r}_m + b_{m+r}b'^{2^r}_{m-r} = 0$ which again leads to (3.25).

Equality (3.25) implies $\mu L(x) = \mu b + \text{tr}(\mu b_0 x)$ and then, because of (3.24), we get

$$\mu[L(x) + L'(F(x))] = \mu b' + \mu b + \text{tr}(\mu b_0 x + \mu b'_0 F(x)).$$

Obviously the function $L(x) + L'(F(x))$ cannot be a permutation. Therefore, $L' = $ const and then $L \neq$ const. In this case (3.13) implies

$$L''(x) + \sum_{m \in \mathbb{Z}/n\mathbb{Z}} b'''_m x^{2^m(2^s+1)} = \sum_{m,k \in \mathbb{Z}/n\mathbb{Z}} b_m b^{2^r}_k x^{2^m+2^{k+r}} + L_0(x) \tag{3.26}$$

for some affine function L_0. For some $m \in \mathbb{Z}/n\mathbb{Z}$ we have $b_m \neq 0$ and since $s \neq \pm r$ it is not difficult to note that $b_m^{2^r+1} + b_{m+r}b_{m-r}^{2^r} = 0$. Thus $b_{m+r}, b_{m-r} \neq 0$ and because of $\gcd(r, n) = 1$ we derive $b_t \neq 0$ and

$$\lambda' = b_m b_{m-r}^{-2^r} = b_t b_{t-r}^{-2^r}$$

for all $t \in \mathbb{Z}/n\mathbb{Z}$ and some $\lambda' \in \mathbb{F}_{2^n}$. This leads to the equality

$$\mu' L(x) = \mu' b + \text{tr}(\mu' b_0 x)$$

with $\lambda' = \mu'^{2^r-1}$ which shows that L is not a permutation. This contradiction proves CCZ-inequivalence of F and G. $\square$

CCZ-Inequivalence of Gold and Kasami Functions

Theorem 9 [7] *Let* $F(x) = x^{2^s+1}$, $K(x) = x^{4^r-2^r+1}$ *and* $\gcd(s, n) = \gcd(r, n) = 1$, $1 \leq s < \frac{n}{2}$, $2 \leq r < \frac{n}{2}$.

1) If $3r \neq \pm 1 \bmod n$ *then* F *and* K *are CCZ-inequivalent on* $\mathbb{F}_{2^n}$.
2) If n *is odd and functions* F *and* K^{-1} *are EA-equivalent on* $\mathbb{F}_{2^n}$ *then,* $s = 1$ *and* $3r = \pm 1 \bmod n$.

Proof Let the functions K and F be CCZ-equivalent on $\mathbb{F}_{2^n}$ and let

$$G'(x) = x^{2^{3r}+1},$$
$$G(x) = x^{2^r+1}.$$

Then, there exists an affine automorphism $\mathcal{L} = (L_1, L_2)$ of $\mathbb{F}_{2^n} \times \mathbb{F}_{2^n}$ such that

$$L_2(x, K(x)) = F(L_1(x, K(x))),$$

which implies, by composition by G,

$$L_2(G(x), G'(x)) = F(L_1(G(x), G'(x))),$$

that is, using the notations (3.15–3.21)

$$0 = L''(G(x)) + L'''(G'(x)) + F[L(G(x)) + L'(G'(x))]$$

$$= \left(L(x^{2^r+1}) + L'(x^{2^{3r}+1}) \right)^{2^s+1} + Q(x)$$

$$= Q(x) + \left(c + \sum_{m \in \mathbb{Z}/n\mathbb{Z}} b_m x^{2^m(2^r+1)} + \sum_{m \in \mathbb{Z}/n\mathbb{Z}} b'_m x^{2^m(2^{3r}+1)} \right)$$

$$\times \left(c^{2^s} + \sum_{m \in \mathbb{Z}/n\mathbb{Z}} b_m^{2^s} x^{2^{m+s}(2^r+1)} + \sum_{m \in \mathbb{Z}/n\mathbb{Z}} b'^{2^s}_m x^{2^{m+s}(2^{3r}+1)} \right)$$

$$= Q'(x) + \sum_{m,k \in \mathbb{Z}/n\mathbb{Z}} b_m b_k^{2^s} x^{2^m(2^r+1)+2^{k+s}(2^r+1)}$$

$$+ \sum_{m,k \in \mathbb{Z}/n\mathbb{Z}} b_m b_k'^{2^s} x^{2^m(2^r+1)+2^{k+s}(2^{3r}+1)}$$

$$+ \sum_{m,k \in \mathbb{Z}/n\mathbb{Z}} b_m' b_k^{2^s} x^{2^m(2^{3r}+1)+2^{k+s}(2^r+1)}$$

$$+ \sum_{m,k \in \mathbb{Z}/n\mathbb{Z}} b_m' b_k'^{2^s} x^{2^m(2^{3r}+1)+2^{k+s}(2^{3r}+1)},$$

where Q and Q' are quadratic.

Suppose that L and L' are not constant. Then $b_m, b_k' \neq 0$ for some $m, k \in \mathbb{Z}/n\mathbb{Z}$.

In the expression above we consider the items with the exponent $2^m(2^r + 1) + 2^{k+s}(2^{3r}+1)$. Since $\gcd(r,n) = 1, r \neq \pm 1$, and if $3r \neq \pm 1$, then it is easy but tedious to check that this exponent has the 2-weight at least 3 and differs from exponents of the types $2^t(2^{3r} + 1) + 2^p(2^{3r} + 1)$ and $2^t(2^r + 1) + 2^p(2^r + 1)$. Thus we get the equality $b_m b_k'^{2^s} + b_{k+s}' b_{m-s}^{2^s} = 0$. Since $b_m, b_k' \neq 0$ then $b_{k+s}', b_{m-s} \neq 0$ and

$$b_m b_{m-s}^{-2^s} = b_{k+s}' b_k'^{-2^s}.$$

Repeating this step for $k + s, k + 2s,...,$ instead of k and for $m - s, m - 2s,...,$ instead of m, because of $\gcd(s,n) = 1$ we get

$$\lambda = b_m b_{m-s}^{-2^s} = b_{k+s}' b_k'^{-2^s} \tag{3.27}$$

for all $m, k \in \mathbb{Z}/n\mathbb{Z}$.

Like in the proof of Theorem 8 from the equality (3.27) we get

$$\mu[L(x) + L'(K(x))] = \mu b' + \mu b + \mathrm{tr}(\mu b_0 x + \mu b_0' K(x)),$$

where $\lambda = \mu^{2^s-1}$. Thus $L_1(x, K(x))$ is not a permutation, a contradiction. Therefore, L or L' is constant and F is then EA-equivalent to K or to the inverse of K. We know that F and K are not EA-equivalent because algebraic degree of K is $r + 1$ while F is quadratic. Thus we need to consider only the case $L = \mathrm{const}$ and $L' \neq \mathrm{const}$. We have $b_m' \neq 0$ for some m and $2^m(2^{3r} + 1) + 2^{m+s}(2^{3r} + 1)$ has 2-weight at least 3 except the cases when $s = \pm 1, 3r = \pm 1$. With the same arguments as above we get that $L_1(x, K(x))$ is not a permutation. If $s = \pm 1$ and $3r = \pm 1$ then the inverse of K (if it exists) may be EA-equivalent to F in some cases. For instance, $K^{-1} = F^4$ for $s = 1, r = 2, n = 5$.   □

Remark 4   It is proven in [13] that the Kasami functions over $\mathbb{F}_{2^7}$ are CCZ-inequivalent to any quadratic function.

CCZ-Inequivalence of Gold and Welch Functions

Theorem 10 [7] *Let $F(x) = x^{2^s+1}$ and $G(x) = x^{2^t+3}$ with $\gcd(s,n) = 1$, $1 \leq s \leq \frac{n-1}{2}$, $t = \frac{n-1}{2} \geq 4$. Then F and G are CCZ-inequivalent on $\mathbb{F}_{2^n}$.*

Proof Let the functions F and G be CCZ-equivalent on $\mathbb{F}_{2^n}$. Then, there exists an affine automorphism $\mathcal{L} = (L_1, L_2)$ of $\mathbb{F}_{2^n} \times \mathbb{F}_{2^n}$ such that

$$L_2(x, G(x)) = F(L_1(x, G(x))),$$

and $L_1(x, G(x))$ a permutation. Using the notations (3.15–3.21) we get

$$0 = L''(x) + L'''(G(x)) + F[L(x) + L'(G(x))]$$

$$= Q(x) + \left(c + \sum_{m \in \mathbb{Z}/n\mathbb{Z}} b_m x^{2^m} + \sum_{m \in \mathbb{Z}/n\mathbb{Z}} b'_m x^{2^m(2^t+3)} \right)^{2^s+1}$$

$$= Q(x) + \left(c + \sum_{m \in \mathbb{Z}/n\mathbb{Z}} b_m x^{2^m} + \sum_{m \in \mathbb{Z}/n\mathbb{Z}} b'_m x^{2^m(2^t+3)} \right)$$

$$\times \left(c^{2^s} + \sum_{m \in \mathbb{Z}/n\mathbb{Z}} b_m^{2^s} x^{2^{m+s}} + \sum_{m \in \mathbb{Z}/n\mathbb{Z}} b'^{2^s}_m x^{2^{m+s}(2^t+3)} \right)$$

$$= Q'(x) + \sum_{m,k \in \mathbb{Z}/n\mathbb{Z}} b'_m b'^{2^s}_k x^{2^m(2^t+3)+2^{k+s}(2^t+3)}$$

$$+ \sum_{m,k \in \mathbb{Z}/n\mathbb{Z}} b_m b'^{2^s}_k x^{2^m+2^{k+s}(2^t+3)}$$

$$+ \sum_{m,k \in \mathbb{Z}/n\mathbb{Z}} b'_m b_k^{2^s} x^{2^m(2^t+3)+2^{k+s}} \tag{3.28}$$

where Q and Q' are cubic.

Since the algebraic degree of G is 3 for $n > 3$ then F and G are EA-inequivalent. Therefore, $L' \neq \text{const}$ and $b'_m \neq 0$ for some m.

Since $t = \frac{n-1}{2} \geq 4$ then $2^m(2^t+3) + 2^{m+s}(2^t+3)$ has 2-weight at least 5 when $s \neq 1, t$. If either $s = 1$ or $s = t$ then $2^m(2^t+3) + 2^{m+s}(2^t+3)$ has 2-weight at least 4 and it differs from the exponents of the items in the first and second sums in (3.28). The equality (3.28) implies $b'^{2^s+1}_m = b'_{m+s} b'^{2^s}_{m-s}$ and then $b'_{m+s}, b'_{m-s} \neq 0$. Because of $\gcd(n,s) = 1$ we get $b'_t \neq 0$ and

$$\lambda = b'_{m+s} b'^{-2^s}_m = b'_t b'^{-2^s}_{t-s} \tag{3.29}$$

for any t.

For $m \neq k+s$ the items $b'_m b'^{2^s}_k x^{2^m(2^t+3)+2^{k+s}(2^t+3)}$ and $b'_{k+s} b'^{2^s}_{m-s} x^{2^m(2^t+3)+2^{k+s}(2^t+3)}$ differ and cancel pairwise because of (3.29). In the case $m = k + s$ the sum gives items with the exponents of 2-weight not greater than 3.

Because of (3.29) we get

$$\mu L'(x) = \mu b' + \text{tr}(\mu b'_0 x), \tag{3.30}$$

where $\lambda = \mu^{2^s-1}$. Therefore, L' is not a permutation and then $L \neq$ const. We have $b_m \neq 0$ for some m and considering the items with the exponent $2^m + 2^{m+s}(2^t + 3)$ of 2-weight 4 we get $b_m b'^{2^s}_m = b'_{m+s} b^{2^s}_{m-s}$ and $b_{m-s} \neq 0$. This leads to the equality

$$b_t b^{-2^s}_{t-s} = b'_{m+s} b'^{-2^s}_m = \lambda$$

for any t, and this, together with (3.30), gives

$$\mu[L(x) + L'(G(x))] = \mu b' + \mu b + \text{tr}(\mu b_0 x + \mu b'_0 G(x)).$$

Thus, $L(x) + L'(G(x))$ is not a permutation. This contradiction shows that F and G are CCZ-inequivalent. $\square$

Remark 5 It was checked with a computer that if $1 < t < 4$ then F is EA-equivalent to G^{-1} only in case $n = 5$, $s = 2$.

Inverse and Dobbertin Functions

Proposition 9 [7] *When $n \geq 5$ the inverse and Dobbertin functions are CCZ-inequivalent to other functions of Table 2.2.*

Proof Gold, Kasami, Welch and Niho functions have the following property: the number $2^{\lfloor \frac{n+1}{2} \rfloor}$ divides all the values in their Walsh spectra (see [8, 9, 12, 14, 15, 17]). It follows from [16] that the inverse function does not have this property. Besides, it is proven in [9] that $2^{\frac{2n}{5}+1}$ cannot be a divisor of all the values in the Walsh spectrum of a Dobbertin function. Since the extended Walsh spectrum of a function is invariant under CCZ-equivalence then Gold, Kasami, Welch and Niho functions are CCZ-inequivalent to the inverse and Dobbertin functions. $\square$

In general the inverse and Dobbertin APN functions are CCZ-inequivalent. Indeed, computer experiments show that the extended Walsh spectra of the inverse and Dobbertin APN functions differ (when $n \neq 5$). Besides, Dobbertin functions are APN also for n even while the inverse function is APN only for n odd.

References

1. L. Budaghyan and C. Carlet. On CCZ-equivalence and its use in secondary constructions of bent functions. *Preproceedings of International Workshop on Coding and Cryptography WCC 2009*, pp. 19–36, 2009.
2. L. Budaghyan and C. Carlet. CCZ-equivalence of single and multi output Boolean functions. *Post-proceedings of the 9-th International Conference on Finite Fields and Their Applications Fq'09*, Contemporary Math., AMS, v. 518, pp. 43–54, 2010.
3. L. Budaghyan and T. Helleseth. New perfect nonlinear multinomials over $F_{p^{2k}}$ for any odd prime p. *Proceedings of the International Conference on Sequences and Their Applications SETA 2008*, Lecture Notes in Computer Science 5203, pp. 403–414, Lexington, USA, Sep. 2008.
4. L. Budaghyan and T. Helleseth. On Isotopisms of Commutative Presemifields and CCZ-Equivalence of Functions. Special Issue on Cryptography of *International Journal of Foundations of Computer Science*, v. 22(6), pp. 1243–1258, 2011. Preprint at http://eprint.iacr.org/2010/507
5. L. Budaghyan and T. Helleseth. New commutative semifields defined by new PN multinomials. *Cryptography and Communications: Discrete Structures, Boolean Functions and Sequences*, v. 3(1), pp. 1–16, 2011.
6. L. Budaghyan, C. Carlet, A. Pott. New Classes of Almost Bent and Almost Perfect Nonlinear Functions. *IEEE Trans. Inform. Theory*, vol. 52, no. 3, pp. 1141–1152, March 2006.
7. L. Budaghyan, C. Carlet, G. Leander. On inequivalence between known power APN functions. *Proceedings of the International Workshop on Boolean Functions: Cryptography and Applications, BFCA 2008*, Copenhagen, Denmark, May 2008.
8. A. Canteaut, P. Charpin, H. Dobbertin. Weight divisibility of cyclic codes, highly nonlinear functions on F_{2^m}, and crosscorrelation of maximum-length sequences. *SIAM Journal on Discrete Mathematics*, 13(1), pp. 105–138, 2000.
9. A. Canteaut, P. Charpin, and H. Dobbertin. Binary m-sequences with three-valued crosscorrelation: A proof of Welch's conjecture. *IEEE Trans. Inform. Theory*, 46 (1), pp. 4–8, 2000.
10. C. Carlet, P. Charpin and V. Zinoviev. Codes, bent functions and permutations suitable for DES-like cryptosystems. *Designs, Codes and Cryptography*, 15(2), pp. 125–156, 1998.
11. P. Charpin, G. Kyureghyan. On a class of permutation polynomials over F_{2^n}. *Proceedings of SETA 2008, Lecture Notes in Computer Science* 5203, pp. 368–376, 2008.
12. R. Gold. Maximal recursive sequences with 3-valued recursive crosscorrelation functions. *IEEE Trans. Inform. Theory*, 14, pp. 154–156, 1968.
13. F. Göloğlu and A. Pott. Almost perfect nonlinear functions: a possible geometric approach. *Proceedings of Academic Contact Forum: Coding Theory and Cryptography*, Brussels, 2007.
14. H. Hollmann and Q. Xiang. A proof of the Welch and Niho conjectures on crosscorrelations of binary m-sequences. *Finite Fields and Their Applications 7*, pp. 253–286, 2001.
15. T. Kasami. The weight enumerators for several classes of subcodes of the second order binary Reed-Muller codes. *Inform. and Control*, 18, pp. 369–394, 1971.
16. G. Lachaud and J. Wolfmann. The Weights of the Orthogonals of the Extended Quadratic Binary Goppa Codes. *IEEE Trans. Inform. Theory*, vol. 36, pp. 686–692, 1990.
17. K. Nyberg. Differentially uniform mappings for cryptography. *Advances in Cryptography, EUROCRYPT'93*, Lecture Notes in Computer Science 765, pp. 55–64, 1994.
18. A. Pott, Y. Zhou. CCZ and EA equivalence between mappings over finite Abelian groups. *Des. Codes Cryptography* 66(1–3), pp. 99–109, 2013.

Chapter 4
Bent Functions

4.1 On the Structure of This Chapter

This chapter is dedicated to our results on analysis and construction of bent functions [2, 7–9].

As observed in the previous chapter, CCZ-equivalence of bent vectorial functions over $\mathbb{F}_2^n$ reduces to their EA-equivalence. In Sect. 4.2 we show that in spite of this fact, CCZ-equivalence can be used for constructing bent functions which are new up to EA-equivalence and therefore to CCZ-equivalence: applying CCZ-equivalence to a non-bent vectorial function F which has some bent components, we get a function F which also has some bent components and whose bent components are CCZ-inequivalent to the components of the original function F. Using this approach we construct classes of nonquadratic bent Boolean and bent vectorial functions [2].

Section 4.3 is dedicated to a problem raised in [13]. In 1998, Carlet, Charpin and Zinoviev characterized APN and AB (n, n)-functions by means of associated $2n$-variable Boolean functions. In particular, they proved that a function F is AB if and only if the associated Boolean function γ_F is bent. This observation leads to potentially new bent functions associated to the known AB functions, or at least gives new insight on known bent functions. However, representations of γ_F were known only for Gold AB power functions and determining γ_F for the rest of AB functions was an open problem. We determine γ_F for most of the known families of APN and AB functions [7].

In Sect. 4.4 we solve a problem which dates back to 1974. In his thesis [16], Dillon introduced a family of bent functions denoted by H, where bentness is proven under some conditions which were not obvious to achieve. In this class, Dillon was able to exhibit only functions belonging to the completed Maiorana-McFarland class. In [12] it was observed that the completed class of H contains all bent functions of the, so called, Niho type which were introduced in [20] by Dobbertin et al. We prove that two classes of binomial Niho bent functions do not belong to the completed MM class. This implies that the class H contains functions which do not belong to the

© Springer International Publishing Switzerland 2014

L. Budaghyan, *Construction and Analysis of Cryptographic Functions*,
DOI 10.1007/978-3-319-12991-4_4

completed Maiorana-McFarland class and, therefore, the class H is not contained in the completed MM class [9].

In the last section of this chapter we study the relation between the generalized bent functions of Table 2.1 and completed class of Maiorana-McFarland functions. In the binary case, the completed MM class contains all quadratic bent functions which are the simplest and best understood. However, this does not hold in the generalized case. First, for p odd there exist quadratic bent functions over $\mathbb{F}_{p^n}$ when n is odd while Maiorana-McFarland bent functions are defined only for n even. For the case n even, we provide examples of quadratic generalized bent functions not belonging to the completed MM class. Moreover, we prove that almost all of the known classes do not intersect with the completed MM class. This leads us to the conclusion that in general, the Maiorana-McFarland construction is less overall than in the binary case even for the case n even [8].

4.2 Constructing New Bent Boolean Functions Using CCZ-Equivalence

We show now that, despite the result of the previous section, CCZ-equivalence can be used for constructing new bent Boolean functions, by applying it to non-bent vectorial functions which admit bent components. We give two examples illustrating this fact. Starting from Gold functions and considering two classes of APN functions which have been shown CCZ-equivalent to them, we derived two infinite classes of bent Boolean functions which are CCZ-inequivalent to the bent components of the Gold functions, and we also deduced new families of vectorial bent functions.

Let i be a positive integer. Let us define for n even the (n, n)-function:

$$F(x) = x^{2^i+1} + (x^{2^i} + x + 1)\mathrm{tr}_n(x^{2^i+1}), \tag{4.1}$$

and for n divisible by 6 the (n, n)-function:

$$G(x) = \left(x + \mathrm{tr}_n^3\left(x^{2(2^i+1)} + x^{4(2^i+1)} \right) \right.$$
$$\left. + \mathrm{tr}_n(x)\mathrm{tr}_n^3\left(x^{2^i+1} + x^{2^{2i}(2^i+1)} \right) \right)^{2^i+1}. \tag{4.2}$$

The functions F and G correspond to the first (under condition of n being even) and the second cases in Table 2.5. They were constructed in [3] by applying CCZ-equivalence to the Gold function $F'(x) = x^{2^i+1}$. When $\gcd(i, n) = 1$ these functions are APN, the function F has algebraic degree 3 (for $n \geq 4$), and the function G has algebraic degree 4 (however, some components of F and G have lower algebraic degrees) [3]. Since the algebraic degrees of non-affine functions are preserved by EA-equivalence, then F and G are EA-inequivalent to F'. We know (see e.g. [22]) that if $n/\gcd(n, i)$ is even and $b \in \mathbb{F}_{2^n}$ is not the $(2^i + 1)$-th power of an element of $\mathbb{F}_{2^n}$, then the Boolean function $\mathrm{tr}_n(bF'(x))$ is bent. In general, if a vectorial function

H has some bent components, this does not yet imply that a function CCZ-equivalent to H has necessarily bent components. Below we show that the two classes (4.1) and (4.2) above have bent nonquadratic components which are CCZ-inequivalent to the components of F'.

4.2.1 The Infinite Class of the Functions F

Let us determine the bent cubic components of function (4.1).

Theorem 11 [2] *Let $n \geq 6$ be an even integer and i be a positive integer not divisible by $n/2$ such that $n/\gcd(i, n)$ is even. Let the function F be given by (4.1), and $b \in \mathbb{F}_{2^n} \setminus \mathbb{F}_{2^i}$. Then the Boolean function $f_b(x) = \mathrm{tr}_n(bF(x))$ has algebraic degree 3, and it is bent if and only if neither b nor $b + 1$ are the $(2^i + 1)$-th powers of elements of $\mathbb{F}_{2^n}$.*

Proof Firstly we prove that for $n/\gcd(i, n)$ even and $b \in \mathbb{F}_{2^n}$ the function f_b is bent if and only if neither b nor $b + 1$ is the $(2^i + 1)$-th power of an element of $\mathbb{F}_{2^n}$.

By Theorem 2 of [3], which proves that the function F is CCZ-equivalent to $F'(x) = x^{2^i+1}$, the graph of F' is mapped to the graph of F by the linear involution

$$\mathcal{L}(x, y) = (L_1(x, y), L_2(x, y)) = (x + \mathrm{tr}_n(y), y).$$

It is shown in the proof of Proposition 2 of [3] (and straightforward to check) that for any $a, b \in \mathbb{F}_{2^n}$

$$\lambda_{F'}(a, b) = \lambda_F(\mathcal{L}^{-1*}(a, b)), \tag{4.3}$$

where $\mathcal{L}^{-1*}$ is the adjoint operator of $\mathcal{L}^{-1}$, that is, for any $(x, y), (x', y') \in \mathbb{F}_{2^n}^2$:

$$(x, y) \cdot \mathcal{L}^{-1*}(x', y') = \mathcal{L}^{-1}(x, y) \cdot (x', y'),$$

where $(x, y) \cdot (x', y') = \mathrm{tr}_n(xx') + \mathrm{tr}_n(yy')$.
The adjoint operator of $\mathcal{L}^{-1} = \mathcal{L}$ is

$$\mathcal{L}^*(x, y) = (L_1^*(x, y), L_2^*(x, y)) = (x, y + \mathrm{tr}_n(x)). \tag{4.4}$$

Indeed,

$$\begin{aligned}
\mathcal{L}(x, y) \cdot (x', y') &= \mathrm{tr}_n\left((x + \mathrm{tr}_n(y))x'\right) + \mathrm{tr}_n(yy') \\
&= \mathrm{tr}_n(xx') + \mathrm{tr}_n(y)\mathrm{tr}_n(x') + \mathrm{tr}_n(yy') \\
&= \mathrm{tr}_n(xx') + \mathrm{tr}_n\left(y(y' + \mathrm{tr}_n(x'))\right) \\
&= (x, y) \cdot \mathcal{L}^*(x', y').
\end{aligned}$$

According to (4.3) and (4.4)

$$\lambda_{F'}(a, b) = \lambda_F(a, b + \text{tr}_n(a)),$$

or, equivalently,

$$\lambda_F(a, b) = \lambda_{F'}(a, b + \text{tr}_n(a)).$$

When $n/\gcd(i, n)$ is even, it is known that $\lambda_{F'}(a, b + \text{tr}_n(a)) = \pm 2^{n/2}$ if and only if $b + \text{tr}_n(a)$ is not the $(2^i + 1)$-th power of an element of $\mathbb{F}_{2^n}$ (see e.g. [22]). Hence, f_b is bent if and only if neither b nor $b + 1$ is the $(2^i + 1)$-th power of an element of $\mathbb{F}_{2^n}$.

Now we prove that for $n \geq 6$ and i not divisible by $n/2$ and $b \notin \mathbb{F}_{2^i}$ the function f_b has algebraic degree 3.

Note that $c = b^{2^{n-i}} + b \neq 0$ since $b \notin \mathbb{F}_{2^i}$, and

$$f_b(x) = \text{tr}_n(bx^{2^i+1}) + \text{tr}_n(b(x^{2^i} + x + 1))\text{tr}_n(x^{2^i+1})$$

$$= \text{tr}_n(bx^{2^i+1}) + \text{tr}_n(b)\text{tr}_n(x^{2^i+1}) + \text{tr}_n((b^{2^{n-i}} + b)x)\text{tr}_n(x^{2^i+1})$$

$$= Q(x) + \text{tr}_n(cx)\text{tr}_n(x^{2^i+1}),$$

where Q is quadratic. To prove that f_b is cubic we need to show that there are cubic terms in $\text{tr}_n(cx)\text{tr}_n(x^{2^i+1})$ which do not vanish.

All items in $\text{tr}_n(x^{2^i+1}) = \sum_{j=0}^{n-1} x^{2^{i+j}+2^j}$ are pairwise different since i is not divisible by $n/2$. Indeed, if for some $0 \leq j, k < n$, $k \neq j$, we have $2^{i+j} + 2^j = 2^{i+k} + 2^k \mod (2^n - 1)$ or, equivalently, $i + j = k \mod n$ and $i + k = j \mod n$ then obviously i is divisible by $n/2$.

Let us denote $A_j = \{j - i, j, j + i, j + 2i\}$. Then, since

$$\sum_{0 \leq j < n} c^{2^{j+2i}} x^{2^j + 2^{j+i} + 2^{j+2i}} = \sum_{0 \leq j < n} c^{2^{j+i}} x^{2^{j-i} + 2^j + 2^{j+i}},$$

we have

$$\text{tr}_n(cx)\text{tr}_n(x^{2^i+1}) = \left(\sum_{0 \leq k < n} c^{2^k} x^{2^k} \right) \left(\sum_{0 \leq j < n} x^{2^j + 2^{j+i}} \right)$$

$$= \sum_{0 \leq j < n} c^{2^j} x^{2^{j+1}+2^{j+i}} + \sum_{0 \leq j < n} c^{2^{j+i}} x^{2^j + 2^{j+i+1}}$$

$$+ \sum_{0 \leq j < n} (c^{2^{j-i}} + c^{2^{j+i}}) x^{2^{j-i} + 2^j + 2^{j+i}}$$

$$+ \sum_{\substack{0 \leq j, k < n \\ k \notin A_j}} c^{2^k} x^{2^k + 2^j + 2^{j+i}}.$$

For $n > 4$ all exponents $2^k + 2^j + 2^{j+i}$ in the sum

$$\sum_{\substack{0 \le j,k < n \\ k \notin A_j}} c^{2^k} x^{2^k + 2^j + 2^{j+i}}$$

are pairwise different, have 2-weight 3 and they obviously differ from the exponents in the first three sums above. Hence, the items with these exponents do not vanish and, therefore, f_b has algebraic degree 3. $\qquad\square$

Since F' is quadratic, then according to Corollary 3, the bent nonquadratic components of F are CCZ-inequivalent to the components of F'.

Corollary 8 [2] *The functions f_b of Theorem 11 are CCZ-inequivalent to any component of $F'(x) = x^{2^i+1}$.*

Remark 6 [2] Knowing the number of EA-inequivalent bent components of a given function W we cannot predict how many bent components can have the function W' which is CCZ-equivalent to W. For instance, x^3 has only one bent component up to EA-equivalence while for small values of n we can check that F has at least 2 bent components up to EA-equivalence. Another interesting example is Dillon-Wolfe function, it is CCZ-equivalent to a function with bent components but it itself has no bent component at all [1].

The Existence of Elements b Satisfying the Conditions of Theorem 11 We first show that there always exist elements b satisfying the conditions of Theorem 11. This result is only an existence result. We shall need a more effective one, for building new bent vectorial functions. So, we subsequently point out explicit values of such elements b, under some conditions.

Proposition 10 [2] *Let $n \ge 6$ be an even integer and i be a positive integer not divisible by $n/2$ such that $n/\gcd(i, n)$ is even. There exist at least $\frac{1}{3}(2^n - 1) - 2^{n/2} > 0$ elements $b \in \mathbb{F}_{2^n} \setminus \mathbb{F}_{2^i}$ such that neither b nor $b + 1$ are the $(2^i + 1)$-th powers of elements of $\mathbb{F}_{2^n}$.*

Proof Since $n/\gcd(i, n)$ is even, we have $\gcd(2i, n) = 2\gcd(i, n)$ and we deduce that $\gcd(2^n - 1, 2^{2i} - 1) = 2^{\gcd(2i,n)} - 1 = (2^{\gcd(i,n)} + 1)(2^{\gcd(i,n)} - 1) = (2^{\gcd(i,n)} + 1)$ $\gcd(2^n - 1, 2^i - 1)$. This implies $\gcd(2^n - 1, 2^i + 1) \ge 2^{\gcd(i,n)} + 1 \ge 3$ (note that this bound is tight since if $\gcd(i, n) = 1$ then $\gcd(2^n - 1, 2^i + 1) = 3$). Then the size of the set E of all $(2^i + 1)$-th powers of elements of $\mathbb{F}_{2^n}^*$ is at most $(2^n - 1)/3$ and this implies that $(\mathbb{F}_{2^n} \cap \mathbb{F}_{2^i}) \cup E \cup (1 + E)$ has size at most $2^{n/2} + 2(2^n - 1)/3 < 2^n - 1$ (since $n > 2$). This completes the proof. $\qquad\square$

In the proposition below, we describe some cases where elements b satisfying the conditions of Theorem 11 can be very easily chosen.

Proposition 11 [2] *Let $n \ge 6$ be an even integer, i a positive integer not divisible by $n/2$, and s a divisor of i such that i/s is odd and $\gcd(n, 2s(2^s + 1)) = 2s$. If $b \in \mathbb{F}_{2^{2s}} \setminus \mathbb{F}_{2^s}$ and the function F is given by (4.1) then the Boolean function $f_b(x) = \mathrm{tr}_n(bF(x))$ is bent and has algebraic degree 3.*

Proof We are going to show that under the assumption of this proposition the conditions of Theorem 11 are satisfied. Since n is divisible by $2s$ and i/s is odd then $n/\gcd(i,n)$ is even. We have $b \notin \mathbb{F}_{2^i}$ because $b \in \mathbb{F}_{2^{2s}} \setminus \mathbb{F}_{2^s}$ and i/s is odd. Besides, obviously, $b + 1 \in \mathbb{F}_{2^{2s}} \setminus \mathbb{F}_{2^s}$. Hence, we need only to prove that any element b in $\mathbb{F}_{2^{2s}} \setminus \mathbb{F}_{2^s}$ is not the $(2^i + 1)$-th power of an element of $\mathbb{F}_{2^n}$.

Note that if the element b is not the $(2^s + 1)$-th power of an element of $\mathbb{F}_{2^n}$ then it is not the $(2^i + 1)$-th power of an element of $\mathbb{F}_{2^n}$. Indeed, for any positive integer u and any positive odd integer v the number $2^{uv} + 1$ is divisible by $2^u + 1$ since

$$2^{uv} + 1 = 2^u + 1 + (2^{2u} - 1)(2^u + 2^{3u} + 2^{5u} + .. + 2^{u(v-2)}), \qquad (4.5)$$

and, therefore, recalling that i/s is odd, $2^s + 1$ is a divisor of $2^i + 1$.

Since $b \in \mathbb{F}_{2^{2s}} \setminus \mathbb{F}_{2^s}$ then there exists a primitive element α of $\mathbb{F}_{2^n}^*$, and a positive integer k not divisible by $2^s + 1$, such that $b = \alpha^{k(2^n - 1)/(2^{2s} - 1)}$. Obviously, b is the $(2^s + 1)$-th power of an element of $\mathbb{F}_{2^n}$ if and only if k is divisible by $r = (2^s + 1)/\gcd(2^s + 1, (2^n - 1)/(2^{2s} - 1))$. Hence, if we can prove that $r = 2^s + 1$, that is, $2^n - 1$ is not divisible by $(2^s + 1)q$ for any divisor $q \neq 1$ of $2^s + 1$, then b is not the $(2^s + 1)$-th power of an element of $\mathbb{F}_{2^n}$ (and, therefore, is not the $(2^i + 1)$-th power of an element of $\mathbb{F}_{2^n}$), and by Theorem 11 the function f_b is bent and has algebraic degree 3.

Let $q \neq 1$ be any divisor of $2^s + 1$ and n be divisible by $2s$. Below we prove that $2^n - 1$ is divisible by $(2^s + 1)q$ if and only if n is divisible by $2sq$.

If n is divisible by $2sq$ then $2^n - 1$ is divisible by $2^{2sq} - 1$ and, therefore, by $2^{sq} + 1$. Since q is odd (being a divisor of $2^s + 1$) then using (4.5) we get

$$2^{sq} + 1 = (2^s + 1)\Big(1 + (2^s - 1)(2^s + 2^{3s} + .. + 2^{s(q-2)})\Big)$$

$$= (2^s + 1)\Big(1 + (2^s + 1)(2^s + 2^{3s} + .. + 2^{s(q-2)})$$

$$- 2(2^s + 2^{3s} + .. + 2^{s(q-2)})\Big)$$

$$= (2^s + 1)\Big(1 + (2^s + 1)(2^s + 2^{3s} + .. + 2^{s(q-2)})$$

$$+ (q - 1) - 2\big((2^s + 1) + (2^{3s} + 1) + ... + (2^{s(q-2)} + 1)\big)\Big)$$

$$= (2^s + 1)^2\Big(2^s + 2^{3s} + ... + 2^{s(q-2)}\Big) + (2^s + 1)q$$

$$- 2(2^s + 1)\Big((2^s + 1) + (2^{3s} + 1) + ... + (2^{s(q-2)} + 1)\Big) \qquad (4.6)$$

which is divisible by $(2^s + 1)q$ because q is a divisor of $2^s + 1$ and because for any odd positive integer v the number $2^{sv} + 1$ is divisible by $2^s + 1$ as it is observed above. Hence, $2^{sq} + 1$, and therefore also $2^n - 1$, are divisible by $(2^s + 1)q$.

Let now n be divisible by $2s$ but not by $2sq$. Then there exist positive integers w and t such that $1 \leq t < q$ and $n = 2s(wq + t)$. Then

$$2^n - 1 = 2^{2st}(2^{2swq} - 1) + (2^{2st} - 1). \qquad (4.7)$$

As it is shown above $2^{2swq} - 1$ is divisible by $(2^s + 1)q$ because the number $2swq$ is divisible by $2sq$. Therefore, because of (4.7), the number $2^n - 1$ is divisible by $(2^s + 1)q$ if and only if $2^{2st} - 1$ is divisible by $(2^s + 1)q$. But $2^{2st} - 1$ is not divisible by $(2^s + 1)q$ as we show below by considering separately the cases t odd and t even.

For t odd, using equality (4.6) and remembering that for any positive odd integer v the number $2^{sv} + 1$ is divisible by $2^s + 1$, we get

$$2^{st} + 1 = (2^s + 1)^2 \left(2^s + 2^{3s} + \ldots + 2^{s(t-2)}\right) + (2^s + 1)t$$
$$- 2(2^s + 1)\left((2^s + 1) + (2^{3s} + 1) + \ldots + (2^{s(t-2)} + 1)\right)$$
$$= (2^s + 1)^2 T + (2^s + 1)t$$

for some integer T. Hence, $2^{st} + 1$ is divisible by $2^s + 1$ but not by $(2^s + 1)q$, and, since $2^{st} - 1$ is not divisible by q (otherwise the odd integer q would be a divisor of $2^{st} + 1$ and $2^{st} - 1$ which is obviously impossible), then the number $2^{2st} - 1$ is also divisible by $2^s + 1$ but not by $(2^s + 1)q$.

For t even

$$2^{st} - 1 = (2^{2s} - 1)(1 + 2^{2s} + \ldots + 2^{s(t-2)})$$
$$= (2^{2s} - 1)\left(t/2 + (2^{2s} - 1) + (2^{4s} - 1) + \ldots + (2^{s(t-2)} - 1)\right)$$
$$= (2^{2s} - 1)t/2 + (2^s + 1)^2 R$$

for some integer R. Hence, $2^{st} - 1$ is divisible by $2^s + 1$ but not by $(2^s + 1)q$. The odd integer $q \neq 1$ is a divisor of $2^s + 1$, and therefore it is a divisor of $2^{st} - 1$. Then, obviously, it is not a divisor of $2^{st} + 1 = (2^{st} - 1) + 2$. Thus, $2^{2st} - 1$ cannot be divisible by $(2^s + 1)q$.

Hence, for both t odd and t even the number $2^{2st} - 1$ is not divisible by $(2^s + 1)q$, and, therefore, $2^n - 1$ is not divisible by $(2^s + 1)q$. $\square$

The Relation of the Functions of Theorem 11 to the Maiorana-McFarland Class of Bent Functions Many bent functions found in trace representation recalled in Sect. 2.2.1 (and listed e.g. in [10]) are in the completed Maiorana-McFarland class. It is interesting to see whether this is also the case of the bent functions of Theorem 11. However, it is in general difficult to determine what is the exact intersection between a given infinite class of bent functions and the completed Maiorana-McFarland class. Below we prove a partial result: the functions f_b of Theorem 11 belong to the completed Maiorana-McFarland class when b belongs to $\mathbb{F}_{2^{n/2}}$.

Proposition 12 [2] *The bent functions f_b of Theorem 11 belong to the completed Maiorana-McFarland class when $b \in \mathbb{F}_{2^{n/2}}$. In particular, all the functions of Proposition 11 are in the completed Maiorana-McFarland class when n is divisible by $4s$.*

Proof To check whether f_b is in the Maiorana-McFarland class, we need to see whether there exists an $n/2$-dimensional vector space such that the second order derivatives

$$D_a D_c f_b(x) = f_b(x) + f_b(x + a) + f_b(x + c) + f_b(x + a + c)$$

vanish when a and c belong to this vector space. We have

$$f_b(x) = \mathrm{tr}_n(bx^{2^i+1}) + \mathrm{tr}_n(b(x^{2^i} + x + 1))\,\mathrm{tr}_n(x^{2^i+1}),$$

$$D_a f_b(x) = \mathrm{tr}_n(bx^{2^i+1}) + \mathrm{tr}_n(bx^{2^i+1} + bax^{2^i} + ba^{2^i}x + ba^{2^i+1})$$
$$+ \mathrm{tr}_n(b(x^{2^i} + x + 1))\mathrm{tr}_n(x^{2^i+1})$$
$$+ \mathrm{tr}_n(b(x^{2^i} + x + 1 + a^{2^i} + a))\mathrm{tr}_n(x^{2^i+1} + ax^{2^i} + a^{2^i}x + a^{2^i+1})$$
$$= \mathrm{tr}_n(bax^{2^i} + ba^{2^i}x + ba^{2^i+1}) + \mathrm{tr}_n(b(a^{2^i} + a))\mathrm{tr}_n(x^{2^i+1})$$
$$+ \mathrm{tr}_n(b(x^{2^i} + x + 1))\mathrm{tr}_n(ax^{2^i} + a^{2^i}x + a^{2^i+1})$$
$$+ \mathrm{tr}_n(b(a^{2^i} + a))\mathrm{tr}_n(ax^{2^i} + a^{2^i}x + a^{2^i+1}),$$

$$D_a D_c f_b(x) = \mathrm{tr}_n(bac^{2^i} + ba^{2^i}c) + \mathrm{tr}_n(b(a^{2^i} + a))\mathrm{tr}_n(cx^{2^i} + c^{2^i}x + c^{2^i+1})$$
$$+ \mathrm{tr}_n(b(c^{2^i} + c))\mathrm{tr}_n(ax^{2^i} + a^{2^i}x + a^{2^i+1})$$
$$+ \mathrm{tr}_n(b(x^{2^i} + x + 1))\mathrm{tr}_n(ac^{2^i} + a^{2^i}c)$$
$$+ \mathrm{tr}_n(b(c^{2^i} + c))\mathrm{tr}_n(ac^{2^i} + a^{2^i}c)$$
$$+ \mathrm{tr}_n(b(a^{2^i} + a))\mathrm{tr}_n(ac^{2^i} + a^{2^i}c)$$
$$= \mathrm{tr}_n(\lambda x) + \epsilon,$$

where

$$\lambda = (c^{2^{n-i}} + c^{2^i})\mathrm{tr}_n(b(a^{2^i} + a)) + (a^{2^{n-i}} + a^{2^i})\mathrm{tr}_n(b(c^{2^i} + c))$$
$$+ (b^{2^{n-i}} + b)\mathrm{tr}_n(ac^{2^i} + a^{2^i}c),$$

$$\epsilon = \mathrm{tr}_n(bac^{2^i} + ba^{2^i}c) + \mathrm{tr}_n(b(a^{2^i} + a))\mathrm{tr}_n(c^{2^i+1})$$
$$+ \mathrm{tr}_n(b(c^{2^i} + c))\mathrm{tr}_n(a^{2^i+1}) + \mathrm{tr}_n(b)\mathrm{tr}_n(ac^{2^i} + a^{2^i}c)$$
$$+ \mathrm{tr}_n(b(c^{2^i} + c))\mathrm{tr}_n(ac^{2^i} + a^{2^i}c) + \mathrm{tr}_n(b(a^{2^i} + a))\mathrm{tr}_n(ac^{2^i} + a^{2^i}c).$$

The function $D_a D_c f_b$ is null if and only if $\epsilon = \lambda = 0$. Then the $n/2$-dimensional vector space can be taken equal to $\mathbb{F}_{2^{n/2}}$. Indeed, if $a, b, c \in \mathbb{F}_{2^{n/2}}$, then λ and ϵ are null since the trace of any element of $\mathbb{F}_{2^{n/2}}$ is null. If, under the conditions of Proposition 11, n is divisible by $4s$ then $b \in \mathbb{F}_{2^{2s}} \subset \mathbb{F}_{2^{n/2}}$.   □

Remark 7 [2] For $n \geq 8$ the functions f_b are not in class $PSap$, up to EA-equivalence, because the degree of $PSap$ functions is always $n/2$.

4.2.2 The Infinite Class of the Functions G

We study now the bent components of function (4.2).

Theorem 12 [2] *Let n be a positive integer divisible by 6 and let i be a positive integer not divisible by $n/2$ such that $n/\gcd(i,n)$ is even. Let $b \in \mathbb{F}_{2^n}$ and let G be given by (4.2). Then the Boolean function $g_b(x) = \mathrm{tr}_n(b\,G(x))$ is bent if and only if, for any $d \in \mathbb{F}_8$, the element $b + d + d^2$ is not the $(2^i + 1)$-th power of an element of $\mathbb{F}_{2^n}$. If, in addition, i is divisible by 3 and $b \notin \mathbb{F}_{2^i}$ then g_b has algebraic degree 3. If i is not divisible by 3 then g_b has algebraic degree at least 3, and it is exactly 4 if $n \geq 12$ and either $b \notin \mathbb{F}_8$ or $\mathrm{tr}_3(b) \neq 0$.*

Proof First we are going to prove that for $n/\gcd(i,n)$ even, the function g_b is bent if and only if the element b of $\mathbb{F}_{2^n}$ is such that for any $d \in \mathbb{F}_8$, the element $b + d + d^2$ is not the $(2^i + 1)$-th power of an element of $\mathbb{F}_{2^n}$.

By Theorem 3 of [3], which proves that the function G is CCZ-equivalent to $F'(x) = x^{2^i+1}$, the graph of F' is mapped to the graph of G by the linear involution

$$\mathcal{L}(x, y) = (x + \mathrm{tr}_n^3(y^2 + y^4), y).$$

For the adjoint operator $\mathcal{L}^*$ of $\mathcal{L}^*$ we have

$$\mathcal{L}^*(x, y) = (x, y + \mathrm{tr}_n^3(x^2 + x^4))$$

because

$$\mathrm{tr}_n\left(\mathrm{tr}_n^3(y^2 + y^4)x'\right) = \mathrm{tr}_n\left(\sum_{\substack{0 \leq j \leq n-1 \\ \frac{n}{3} \mid j}} x'y^{2^j}\right) = \mathrm{tr}_n\left(\sum_{\substack{0 \leq j \leq n-1 \\ \frac{n}{3} \mid j}} x'^{2^{n-j}}y\right)$$

$$= \mathrm{tr}_n\left(\sum_{\substack{0 \leq j \leq n-1 \\ \frac{n}{3} \mid j}} x'^{2^j}y\right) = \mathrm{tr}_n\left(\mathrm{tr}_n^3(x'^2 + x'^4)y\right).$$

Since $\mathcal{L}$ and $\mathcal{L}^*$ are involutions and since $\lambda_G(a, b) = \lambda_{F'}(\mathcal{L}^{-1*}(a, b))$, then we get

$$\lambda_G(a, b) = \lambda_{F'}(a, b + \mathrm{tr}_n^3(a^2 + a^4)).$$

Thus, g_b is bent if and only if $b + \mathrm{tr}_n^3(a^2 + a^4)$ is not the $(2^i + 1)$-th power of an element of $\mathbb{F}_{2^n}$ for any a. This proves the first part of Theorem 12.

We prove below that the function g_b has algebraic degree 3 when i is divisible by 3 but not by $n/2$ and $b \notin \mathbb{F}_{2^i}$.

Since $\mathrm{tr}_n^3(x^{2^{2i}(2^i+1)}) = \mathrm{tr}_n^3(x^{2^i+1})$ for i divisible by 3 then

$$G(x) = \left(x + \mathrm{tr}_n^3\left(x^{2(2^i+1)} + x^{4(2^i+1)}\right)\right)^{2^i+1}$$

$$= x^{2^i+1} + \mathrm{tr}_n^3\left(x^{2^i+1} + x^{4(2^i+1)}\right) + (x + x^{2^i})\mathrm{tr}_n^3\left(x^{2(2^i+1)} + x^{4(2^i+1)}\right).$$

Clearly, $c = b + b^{2^{n-i}} \neq 0$ because $b \notin \mathbb{F}_{2^i}$, and, since i is not divisible by $n/2$ then all terms in $\mathrm{tr}_n^3(x^{2(2^i+1)} + x^{4(2^i+1)})$ are pairwise different. For some quadratic function Q, we have

$$g_b(x) = Q(x) + \mathrm{tr}_n\left(b(x + x^{2^i})\mathrm{tr}_n^3\left(x^{2(2^i+1)} + x^{4(2^i+1)}\right)\right)$$

$$= Q(x) + \mathrm{tr}_3\left(\mathrm{tr}_n^3(cx)\mathrm{tr}_n^3\left(x^{2(2^i+1)} + x^{4(2^i+1)}\right)\right).$$

and it is not difficult to see that the cubic terms of g_b do not vanish. Indeed,

$$\mathrm{tr}_3\left(\mathrm{tr}_n^3(cx)\mathrm{tr}_n^3\left(x^{2(2^i+1)} + x^{4(2^i+1)}\right)\right) =$$

$$\sum_{j,k=0}^{n/3-3} c^{2^{3k}} x^{2^{3k}+2^{3j+1}+2^{3j+i+1}} + \sum_{j,k=0}^{n/3-3} c^{2^{3k}} x^{2^{3k}+2^{3j+2}+2^{3j+i+2}}$$

$$+ \sum_{j,k=0}^{n/3-3} c^{2^{3k+1}} x^{2^{3k+1}+2^{3j+2}+2^{3j+i+2}} + \sum_{j,k=0}^{n/3-3} c^{2^{3k+1}} x^{2^{3k+1}+2^{3j+3}+2^{3j+i+3}}$$

$$+ \sum_{j,k=0}^{n/3-3} c^{2^{3k+2}} x^{2^{3k+2}+2^{3j+3}+2^{3j+i+3}} + \sum_{j,k=0}^{n/3-3} c^{2^{3k+2}} x^{2^{3k+2}+2^{3j+4}+2^{3j+i+4}}.$$

The item with the exponent $1 + 2^1 + 2^{i+1}$ of x appears only in the first sum above and, obviously, it does not vanish there. As i is divisible by 3 but not by $n/2$ then this exponent has 2-weight 3.

Let now i be not divisible by 3. We are going to prove that in this case the function g_b has algebraic degree at least 3, and it is exactly 4 if $n \geq 12$, and either $b \notin \mathbb{F}_8$ or $\mathrm{tr}_3(b) \neq 0$. For $n = 6$ it is checked with a computer that g_b has algebraic degree at least 3 for any $b \in \mathbb{F}_{2^6}^*$.

Let $n \geq 12$. For simplicity we consider only the case $i = 1$. Denoting $T(x) = \mathrm{tr}_n^3(x^3)$ we get

$$G(x) = C(x) + \mathrm{tr}_3\left(T(x)^3\right) + \mathrm{tr}_n(x)\left(x\left(T(x) + T(x)^2\right) + x^2\left(T(x) + T(x)^4\right)\right),$$

where

$$C(x) = x^3 + T(x) + \mathrm{tr}_n(x)\left(T(x) + T(x)^4\right) + x\left(T(x) + T(x)^4\right) + x^2\left(T(x)^2 + T(x)^4\right)$$

is a cubic function. Hence,

$$g_b(x) = \mathrm{tr}_n(bC(x)) + \mathrm{tr}_n(b)\mathrm{tr}_3\left(T(x)^3\right)$$

$$+ \mathrm{tr}_n(x)\mathrm{tr}_3\left(T(x)\mathrm{tr}_n^3(bx + bx^2 + (b^2 + b^4)x^4)\right)$$

$$= \mathrm{tr}_n(bC(x)) + \mathrm{tr}_n(b)\left(\sum_{0 \leq j,t < n/3} x^{2^{3j+1}+2^{3j}+2^{3t+2}+2^{3t+1}}\right)$$

$$+ \sum_{0 \leq j,t < n/3} x^{2^{3j+3}+2^{3j+2}+2^{3t+1}+2^{3t}} + \sum_{0 \leq j,t < n/3} x^{2^{3j+3}+2^{3j+2}+2^{3t+2}+2^{3t+1}} \Bigg)$$

$$+ \sum_{\substack{0 \leq j,k < n \\ 0 \leq t < n/3}} u_k x^{2^j+2^k+2^{3t}+2^{3t+1}} + \sum_{\substack{0 \leq j,k < n \\ 0 \leq t < n/3}} v_k x^{2^j+2^k+2^{3t+1}+2^{3t+2}}$$

$$+ \sum_{\substack{0 \leq j,k < n \\ 0 \leq t < n/3}} w_k x^{2^j+2^k+2^{3t+2}+2^{3t+3}}$$

where for $0 \leq k < n$

$$u_k = \begin{cases} b^{2^k} & \text{if } k = 0 \bmod 3 \\ b^{2^{k-1}} & \text{if } k = 1 \bmod 3 \\ (b^2 + b^4)^{2^{k-2}} & \text{if } k = 2 \bmod 3 \end{cases},$$

$$v_k = \begin{cases} b^{2^k} & \text{if } k = 1 \bmod 3 \\ b^{2^{k-1}} & \text{if } k = 2 \bmod 3 \\ (b^2 + b^4)^{2^{k-2}} & \text{if } k = 0 \bmod 3 \end{cases},$$

$$w_k = \begin{cases} b^{2^k} & \text{if } k = 2 \bmod 3 \\ b^{2^{k-1}} & \text{if } k = 0 \bmod 3 \\ (b^2 + b^4)^{2^{k-2}} & \text{if } k = 1 \bmod 3 \end{cases}.$$

The exponent $2^6 + 2^9 + 2^0 + 2^1$ has 2-weight 4 and, obviously, we have items with this exponent only with coefficients u_6 and u_9. Then $u_6 + u_9 = b^{2^6} + b^{2^9} = (b + b^8)^{2^6} \neq 0$ when $b \notin \mathbb{F}_{2^3}$. Hence, in the univariate polynomial representation of g_b the item $x^{2^6+2^9+2^0+2^1}$ has a non-zero coefficient and, therefore, g_b has algebraic degree 4 for $b \notin \mathbb{F}_{2^3}$.

If $b \in \mathbb{F}_{2^3}$ then $\text{tr}_n(b) = 0$. If $\text{tr}_3(b) \neq 0$ then we have items with the exponent $2^6 + 2^8 + 2^0 + 2^1$ only with coefficients u_6 and u_8 and $u_6 + u_8 = b^{2^6} + (b^2 + b^4)^{2^6} = \text{tr}_3(b) \neq 0$. Hence, again g_b has algebraic degree 4 when $b \in \mathbb{F}_{2^3}$ and $\text{tr}_3(b) \neq 0$.

Let $b \in \mathbb{F}_{2^3}$ and $\text{tr}_3(b) = 0$. Then all items with exponents of 2-weight 4 vanish and

$$g_b(x) = \text{tr}_n(bC(x))$$

$$= \text{tr}_n \left(b(x^3 + T(x)) \right) + \text{tr}_3 \left(T(x)\text{tr}_n^3(bx + b^2x^2 + b^2x^4 + b^4x^8) \right)$$

$$= \text{tr}_n \left(b(x^3 + T(x)) \right) + \sum_{\substack{0 \leq k < n \\ 0 \leq t < n/3}} b^2 x^{2^k+2^{3t}+2^{3t+1}}$$

$$+ \sum_{\substack{0 \leq k < n \\ 0 \leq t < n/3}} b^4 x^{2^k+2^{3t+1}+2^{3t+2}} + \sum_{\substack{0 \leq k < n \\ 0 \leq t < n/3}} b x^{2^k+2^{3t+2}+2^{3t+3}}.$$

In g_b, the only item with the exponent $2^0 + 2^1 + 2^3$ has the coefficient b^2. Hence g_b has algebraic degree 3 when $b \in \mathbb{F}_{2^3}^*$ and $\mathrm{tr}_3(b) = 0$.   □

Since F' is quadratic then, according to Corollary 3, the bent nonquadratic components of G are CCZ-inequivalent to the components of F'.

Corollary 9 [2] *The functions g_b of Theorem 12 are CCZ-inequivalent to any component of $F'(x) = x^{2^i+1}$.*

Remark 8 [2] We checked with a computer that for $n = 6$ there are cubic bent components of G which are EA-inequivalent to any component of F. This implies that in general cubic bent components of G are EA-inequivalent to cubic bent components of F.

The Existence of Elements b Satisfying the Conditions of Theorem 12 and Relation to MM Class We prove in Proposition 13 the existence of elements b satisfying the conditions of Theorem 12 for $\gcd(i,n) \neq 1$. The existence of such elements for the case $\gcd(i,n) = 1$ when $\gcd(9,n) \neq 9$ will be shown in Proposition 15.

Proposition 13 [2] *Let n be a positive even integer divisible by 6 and i a positive integer not divisible by $n/2$ such that $n/\gcd(i,n)$ is even and $\gcd(i,n) \neq 1$. There exist at least $\frac{1}{5}(2^n - 1) - 2^{n/2} > 0$ elements $b \in \mathbb{F}_{2^n} \setminus \mathbb{F}_{2^i}$ such that, for any $d \in \mathbb{F}_8$, the element $b + d + d^2$ is not the $(2^i + 1)$-th power of an element of $\mathbb{F}_{2^n}$.*

Proof As in the proof of Proposition 10, we have $\gcd(2^n - 1, 2^i + 1) \geq 2^{\gcd(i,n)} + 1$. This implies $\gcd(2^n - 1, 2^i + 1) \geq 5$. Since the number of $d + d^2$ equals 4 and the size of the set E' of all $(2^i + 1)$-th powers of elements of $\mathbb{F}_{2^n}^*$ is at most $(2^n - 1)/5$, this implies that $(\mathbb{F}_{2^n} \cap \mathbb{F}_{2^i}) \cup (\bigcup_{d \in \mathbb{F}_8} (d + d^2 + E'))$ has size at most $2^{n/2} + 4(2^n - 1)/5 < 2^n - 1$. This completes the proof.   □

Here again, we shall need a more effective result, in order to build a bent vectorial function deduced from G. Next proposition describes cases for i divisible by 3 where elements b satisfying the conditions of Theorem 12 can be very easily chosen.

Proposition 14 [2] *Let i, n, s be positive integers such that i is not divisible by $n/2$, $\gcd(i,6s) = 3s$, and $\gcd(n, 6s(2^{3s} + 1)) = 6s$. If $b \in \mathbb{F}_{2^{6s}} \setminus \mathbb{F}_{2^{3s}}$ and the function G is given by (4.2) then the Boolean function $g_b(x) = \mathrm{tr}_n(bG(x))$ is bent and cubic.*

Proof We are going to show that, under these assumptions, the conditions of Theorem 12 are satisfied. Note that since $\gcd(i,6s) = 3s$ then $\frac{i}{3s}$ is odd, and since $b \in \mathbb{F}_{2^{6s}} \setminus \mathbb{F}_{2^{3s}}$ then $b \notin \mathbb{F}_{2^i}$. Besides, $n/\gcd(i,n)$ is even because $\gcd(i,6s) = 3s$ and $\gcd(n,6s) = 6s$.

According to (4.5) the number $2^i + 1$ is divisible by $2^{3s} + 1$ because $\frac{i}{3s}$ is odd. Therefore if b is not the $(2^{3s} + 1)$-th power of an element of $\mathbb{F}_{2^n}$ then it is not the $(2^i + 1)$-th power of an element of $\mathbb{F}_{2^n}$. Besides, since $b \in \mathbb{F}_{2^{6s}} \setminus \mathbb{F}_{2^{3s}}$ then for any $d \in \mathbb{F}_8$ we have $b + d + d^2 \in \mathbb{F}_{2^{6s}} \setminus \mathbb{F}_{2^{3s}}$. Hence, it is enough to prove that any element b in $\mathbb{F}_{2^{6s}} \setminus \mathbb{F}_{2^{3s}}$ is not the $(2^{3s} + 1)$-th power of an element of $\mathbb{F}_{2^n}$.

Since $b \in \mathbb{F}_{2^{6s}} \setminus \mathbb{F}_{2^{3s}}$ then there exists a primitive element α of $\mathbb{F}_{2^n}$, and a positive integer k not divisible by $2^{3s} + 1$, such that $b = \alpha^{k(2^n - 1)/(2^{6s} - 1)}$. Obviously, b is the $(2^{3s} + 1)$-th power of an element of $\mathbb{F}_{2^n}$ if and only if k is divisible by $r =$

$(2^{3s} + 1)/\gcd(2^{3s} + 1, (2^n - 1)/(2^{6s} - 1))$. But since $\gcd(n, 6s(2^{3s} + 1)) = 6s$ then $r = 2^{3s} + 1$ (see the proof of Proposition 11). Hence, b cannot be the $(2^{3s} + 1)$-th power of an element of $\mathbb{F}_{2^n}$. □

For i not divisible by 3 we obtain a slightly more complex description of some elements b satisfying the conditions of Theorem 12.

Proposition 15 [2] *Let i, n, s be positive integers such that $n \geq 12$, $\gcd(i, 2s) = s$, $\gcd(i, 3) = 1$, $\gcd(n, 6s(2^{3s} + 1)) = 6s$, and the function G be given by (4.2). If $b \in \mathbb{F}_{2^{6s}}$ is such that for any $d \in \mathbb{F}_8$ the element $b + d + d^2$ is not the $(2^s + 1)$-th power of an element of $\mathbb{F}_{2^{6s}}$ then the function $g_b(x) = \operatorname{tr}_n(bG(x))$ is bent and has algebraic degree 4.*

Proof We have that i/s is odd and $n/\gcd(i, n)$ is even because $\gcd(i, 2s) = s$ and $\gcd(n, 6s(2^{3s} + 1)) = 6s$. Then $2^i + 1$ is divisible by $2^s + 1$ due to (4.5). Therefore if b is not the $(2^s + 1)$-th power of an element of $\mathbb{F}_{2^n}$ then it is not the $(2^i + 1)$-th power of an element of $\mathbb{F}_{2^n}$. Besides, since $b \in \mathbb{F}_{2^{6s}}$ then for any $d \in \mathbb{F}_8$ we have $b + d + d^2 \in \mathbb{F}_{2^{6s}}$. Hence we need only to prove that any element $b \in \mathbb{F}_{2^{6s}}$, which is not the $(2^s + 1)$-th power of an element of $\mathbb{F}_{2^{6s}}$, is not the $(2^s + 1)$-th power of an element of $\mathbb{F}_{2^n}$.

Since $b \in \mathbb{F}_{2^{6s}}$ then there exists a primitive element α of $\mathbb{F}_{2^n}$ and a positive integer k such that $b = \alpha^{k(2^n - 1)/(2^{6s} - 1)}$. Since $\gcd(n, 6s(2^{3s} + 1)) = 6s$ then, as shown in the proof of Proposition 11, we have $\gcd(2^{3s} + 1, (2^n - 1)/(2^{6s} - 1)) = 1$, and therefore $\gcd(2^s + 1, (2^n - 1)/(2^{6s} - 1)) = 1$ because $2^s + 1$ is a divisor of $2^{3s} + 1$. Hence b is the $(2^s + 1)$-th power of an element of $\mathbb{F}_{2^n}$ if and only if k is divisible by $2^s + 1$, that is, if and only if b is the $(2^s + 1)$-th power of an element of $\mathbb{F}_{2^{6s}}$. □

For small values of s it is easy to count the exact numbers of elements $b \in \mathbb{F}_{2^{6s}}$ which satisfy the condition of Proposition 15. For instance, for $s = 2$ there are 1736 such elements b, and for $s = 4$ there are 13172960 such elements. For $s = 1$ there are 12 such elements and these elements b are zeros of one of the polynomials $x^6 + x + 1$ and $x^6 + x^4 + x^3 + x + 1$. Hence, if in addition to conditions of Theorem 12 we have $\gcd(i, n) = 1$ and $\gcd(9, n) = 3$ then Proposition 15 ensures the existence of elements satisfying the conditions of this theorem.

Thanks to computer investigations, we know that some of the constructed bent functions g_b (Theorem 12) are neither in MM class nor in PS class:

Proposition 16 [2] *For $n = 12$ and $i = 1$, α a primitive element of $\mathbb{F}_{2^n}$ (determined by MAGMA), the function $\operatorname{tr}_n(\alpha^{19}G(x))$ is a bent function of algebraic degree 4 which is neither in MM class nor in PS class, up to EA-equivalence (that is, up to CCZ-equivalence).*

This shows by an example that having a vectorial function F with bent components all of which are in the MM class, we can construct a function F' CCZ-equivalent to F but which has some non-MM bent components.

Remark 9 [2] For $n \geq 10$ the functions g_b are not in class $PSap$, up to EA-equivalence, because the degree of $PSap$ functions is always $n/2$.

4.2.3 Further Constructions?

Applying CCZ-equivalence to the quadratic APN function $x^3 + tr_n(x^9)$, it is possible to construct classes of nonquadratic APN mappings with some bent components. The same affine transformations $\mathcal{L}$ as those which gave respectively F and G from Gold functions, when they are applied to the graph of $x^3 + tr_n(x^9)$ (which is CCZ-inequivalent to Gold), give graphs of functions as well, and some of the components of the resulting CCZ-equivalent APN functions are bent. Indeed, let n be an even positive integer, $H : \mathbb{F}_{2^n} \to \mathbb{F}_{2^n}$, $H(x) = x^3 + tr_n(x^9)$, then the following functions are CCZ-equivalent to H (see [6])

1) the function with algebraic degree 3

$$x^3 + tr_n(x^9) + (x^2 + x + 1)tr_n(x^3);$$

2) for n divisible by 6 the function with algebraic degree 4

$$\left(x + tr_n^3(x^6 + x^{12}) + tr_n(x)tr_n^3(x^3 + x^{12})\right)^3$$

$$+ tr_n\left(\left(x + tr_n^3(x^6 + x^{12}) + tr_n(x)tr_n^3(x^3 + x^{12})\right)^9\right).$$

The bent components of the functions above have the same algebraic degrees as those of F, resp. G. We could check by a computer that for small values of n the bent components of those functions are equivalent to bent components of F and G, respectively. We do not know if in general the resulting APN functions have bent components inequivalent to those of F and G and it seems difficult to see this mathematically.

For $n = 12$ we give below another example illustrating the application of CCZ-equivalence in constructions of bent functions.

Example 1 [2] Let α be a primitive element of $\mathbb{F}_{2^{12}}$ and $P : \mathbb{F}_{2^{12}} \to \mathbb{F}_{2^{12}}$, $P(x) = \alpha x^3 + \alpha^{256}x^{528} + \alpha^{257}x^{514}$. The function P is EA-equivalent to the trinomial APN function from [2]. Let

$$L_1 = tr_{12}^3(y) + \alpha tr_{12}^3(\alpha^4 x) + \alpha^2 tr_{12}^3(\alpha^{16}x) + \alpha^4 tr_{12}^3(\alpha^{64}x),$$

$$L_2 = tr_{12}^3(x) + \alpha tr_{12}^3(\alpha^4 y) + \alpha^2 tr_{12}^3(\alpha^{16}y) + \alpha^4 tr_{12}^3(\alpha^{64}y).$$

Then the linear function (L_1, L_2) is a permutation of $\mathbb{F}_{2^{12}}^2$ and the function $P_1(x) = L_1(x, P(x))$ is a permutation of $\mathbb{F}_{2^{12}}$. Therefore, the function $P' = P_2 \circ P_1^{-1}$, where $P_2(x) = L_2(x, P(x))$, is CCZ-equivalent to P. The function $tr_{12}(\alpha^9 P(x))$ is bent and has algebraic degree 5, it is EA-inequivalent to any function from MM classes (as checked with a computer). Obviously, it is EA-inequivalent to any bent component of F, F', G, P or any PSap function because of the algebraic degree.

Non-Existence of APN Permutations EA-Equivalent to Functions F and G
Finding APN permutations over $\mathbb{F}_{2^n}$ when n is even is a hard problem. Non-existence of such quadratic functions was proven in [23]. Hence the APN function

$F'(x) = x^{2^i+1}$, $\gcd(i, n) = 1$, n even, is EA-inequivalent to any permutation. However, it is potentially possible that F' is CCZ-equivalent to a nonquadratic APN permutation. For instance, the only known example of an APN permutation for n even is constructed in [1] by applying CCZ-equivalence to a quadratic APN function over $\mathbb{F}_{2^6}$. From this point of view the following facts are interesting.

Corollary 10 [2] *Let n and i be positive integers and $\gcd(i, n) = 1$. If n is even then the APN function F given by (4.1) is EA-inequivalent to any permutation over $\mathbb{F}_{2^n}$. If $\gcd(n, 18) = 6$ then the APN function G given by (4.2) is EA-inequivalent to any permutation over $\mathbb{F}_{2^n}$.*

Proof The function F has bent components by Proposition 10, and G has bent components by Proposition 15. Therefore, F and G are not EA-equivalent to any permutation. □

4.2.4   New Classes of Bent Vectorial Functions in Trace Representation

Let F be a function from $\mathbb{F}_{2^n}$ to itself and $b \in \mathbb{F}_{2^n}^*$. For n divisible by m, the (n, m)-function $\mathrm{tr}_n^m(bF(x))$ is bent if and only if, for any $v \in \mathbb{F}_{2^m}^*$, the Boolean function $\mathrm{tr}_n(bvF(x))$ is bent. Hence we can obtain vectorial bent functions from Theorem 11.

Theorem 13 [2] *Let $n \geq 6$ be an even integer divisible by m and i a positive integer not divisible by $n/2$ and such that $n/\gcd(i, n)$ is even. If $b \in \mathbb{F}_{2^n} \setminus \mathbb{F}_{2^i}$ is such that for any $v \in \mathbb{F}_{2^m}^*$, neither bv nor $bv + 1$ are the $(2^i + 1)$-th powers of elements of $\mathbb{F}_{2^n}$, and the function F is given by (4.1) then the function $f_b(x) = \mathrm{tr}_n^m(bF(x))$ is bent and has algebraic degree 3.*

In particular we obtain the following vectorial bent functions from Proposition 11.

Corollary 11 [2] *Let $n \geq 6$ be an even integer, i a positive integer not divisible by $n/2$ and s a divisor of i such that i/s is odd and $\gcd(n, 2s(2^s + 1)) = 2s$. If $b \in \mathbb{F}_{2^{2s}} \setminus \mathbb{F}_{2^s}$ and the function F is given by (4.1), then the function $f_b(x) = \mathrm{tr}_n^s(bF(x))$ is bent and has algebraic degree 3.*

Proof Since $b \in \mathbb{F}_{2^{2s}} \setminus \mathbb{F}_{2^s}$ then $bv \in \mathbb{F}_{2^{2s}} \setminus \mathbb{F}_{2^s}$ for any $v \in \mathbb{F}_{2^s}^*$. Hence by Proposition 11 the functions $\mathrm{tr}_n(bvF(x))$ are bent and cubic for all $v \in \mathbb{F}_{2^s}^*$, and, therefore, $\mathrm{tr}_n^s(bF(x))$ is bent and has algebraic degree 3. □

Theorem 12 also leads to new bent vectorial functions.

Theorem 14 [2] *Let n be a positive integer divisible by 6, $m > 1$ a divisor of n, and i a positive integer not divisible by $n/2$ such that $n/\gcd(i, n)$ is even. Let $b \in \mathbb{F}_{2^n}$ be such that, for any $d \in \mathbb{F}_8$ and any $v \in \mathbb{F}_{2^m}^*$, $bv + d + d^2$ is not the $(2^i + 1)$-th power of an element of $\mathbb{F}_{2^n}$. If the function G is given by (4.2) then the Boolean function $g_b(x) = \mathrm{tr}_n^m(b\,G(x))$ is bent. If, in addition, i is divisible by 3, and $bv \notin \mathbb{F}_{2^i}$ for some $v \in \mathbb{F}_{2^m}^*$ then g_b has algebraic degree 3. If i is not divisible by 3 then g_b has algebraic* *degree at least* 3, *and it is exactly* 4 *if* $n \geq 12$, *and for some* $v \in \mathbb{F}_{2^m}^*$ *either* $bv \notin \mathbb{F}_8$ *or* $\mathrm{tr}_3(bv) \neq 0$.

Proposition 14 allows us to describe some particular cases of bent vectorial functions of Theorem 14 for i divisible by 3.

Corollary 12 [2] *Let* i, n, s *be positive integers such that* i *is not divisible by* $n/2$, $\gcd(i, 6s) = 3s$, *and* $\gcd(n, 6s(2^{3s} + 1)) = 6s$. *If* $b \in \mathbb{F}_{2^{6s}} \setminus \mathbb{F}_{2^{3s}}$ *and the function* G *is given by* (4.2) *then the function* $g_b(x) = \mathrm{tr}_n^{3s}(bG(x))$ *is bent and cubic*.

Proof Since $b \in \mathbb{F}_{2^{6s}} \setminus \mathbb{F}_{2^{3s}}$ then $bv \in \mathbb{F}_{2^{6s}} \setminus \mathbb{F}_{2^{3s}}$ for any $v \in \mathbb{F}_{2^{3s}}^*$. Hence by Proposition 14 the functions $\mathrm{tr}_n(bvF(x))$ are bent and cubic for all $v \in \mathbb{F}_{2^{3s}}^*$, and, therefore, $\mathrm{tr}_n^{3s}(bF(x))$ is bent and cubic.   □

Next corollary follows from Proposition 15 and refers to the case where i is not divisible by 3.

Corollary 13 [2] *Let* i, n, s *be positive integers such that* $n \geq 12$, $\gcd(i, 2s) = s$, $\gcd(i, 3) = 1$, *and* $\gcd(n, 6s(2^{3s} + 1)) = 6s$. *If the function* G *is given by* (4.2) *and* $b \in \mathbb{F}_{2^{6s}}$ *is such that for any* $d \in \mathbb{F}_8$ *and any* $v \in \mathbb{F}_{2^{3s}}^*$ *the element* $bv + d + d^2$ *is not the* $(2^s + 1)$-*th power in* $\mathbb{F}_{2^{6s}}$ *then the function* $g_b(x) = \mathrm{tr}_n^{3s}(bG(x))$ *is bent and has algebraic degree* 4.

Since $F'(x) = x^{2^i+1}$ is quadratic, then according to Corollary 3:

Corollary 14 [2] *The bent functions* f_b *and* g_b *of Theorems* 13 *and* 14 *(and Corollaries* 11, 12 *and* 13, *in particular) are CCZ-inequivalent to* $\mathrm{tr}_n^m(vF'(x))$ *for any* $v \in \mathbb{F}_{2^n}$ *and any divisor* m *of* n.

Remark 10 [2] To our knowledge there are only three known infinite classes of vectorial bent functions expressed in trace representation $\mathrm{tr}_n^m(F(x))$: the function $\mathrm{tr}_n^m(x^{2^{n/2}+1})$ (which is a Maiorana McFarland function), the function $\mathrm{tr}_n^m(wx^d)$ where n is congruent to 2 mod 4, $d = 2^i + 1$ is a Gold exponent (with $(i, n) = 1$) and w is not a cube, and the function $\mathrm{tr}_n^m(wx^d)$ where n is congruent to 2 mod 4, $d = 4^i - 2^i + 1$ is a Kasami exponent (with $(i, n) = 1$) and w is not a cube (see [16, 22]). The functions we obtain in this section are inequivalent to these functions and so they are new in this sense: Corollary 8 shows that the constructed vectorial bent functions are not CCZ-equivalent to the functions with Gold exponents indicated above; inequivalence to the functions above with Kasami eponents can be easily seen for any n divisible by 4 (because Kasami type bent functions are not defined then) and any n divisible by 6 since the constructed bent vectorial functions have algebraic degree 3 or 4 while for n divisible by 6 there exists no Kasami function of algebraic degree 3 or 4 (Kasami type bent functions have algebraic degree $i + 1$).

4.3   On Bent Functions Associated to AB Functions

In [13], APN and AB functions are characterized by means of associated Boolean functions. For a given function F from $\mathbb{F}_{2^n}$ to itself and for any $a, b \in \mathbb{F}_{2^n}$, define the Boolean function γ_F over $\mathbb{F}_{2^n}^2$ as

$$\gamma_F(a,b) = \begin{cases} 1 & \text{if } a \neq 0 \text{ and } F(x+a) + F(x) = b \text{ has solutions,} \\ 0 & \text{otherwise.} \end{cases}$$

Note that for any power function $F(x) = x^d$ we have $\gamma_F(a,b) = \gamma_F(1, b/a^d)$ for any $a, b \in \mathbb{F}_{2^n}, a \neq 0$.

The theorem below shows connections between properties of F and γ_F.

Theorem 15 [13] *Let F be a function from $\mathbb{F}_{2^n}$ to itself. Then the following properties hold:*

(i) F is APN if and only if γ_F has weight $2^{2n-1} - 2^{n-1}$;
(ii) F is AB if and only if γ_F is bent;
(iii) if F is APN then the function $b \to \gamma_F(a,b)$ is balanced for any $a \neq 0$;
(iv) if F is an APN permutation then the function $a \to \gamma_F(a,b)$ is balanced for any
      $b \neq 0$.

In addition, CCZ-equivalence and EA-equivalence result in the equivalence of the associated functions γ_F.

Proposition 17 [7] *Let F and F' be functions from $\mathbb{F}_{2^n}$ to itself. If F and F' are CCZ-equivalent then there exists an affine permutation $\mathcal{L}$ of $\mathbb{F}_{2^n}^2$ such that*

$$\gamma_{F'} = \gamma_F \circ \mathcal{L}.$$

If F and F' are EA-equivalent then there exist affine permutations A_1, A_2 and an affine function A such that for any $a, b \in \mathbb{F}_2^n$,

$$\gamma_{F'}(a,b) = \gamma_F\Big(A_2(a) + A_2(0),\ A_1^{-1}(A(a) + b + A(0) + A_1(0))\Big).$$

Indeed, if $F' = A_1 \circ F \circ A_2 + A$ then the derivative $D_a F'(x)$ of F' equals $A_1(D_{A_2(a)+A_2(0)}F(A_2(x))) + A(a) + A(0) + A_1(0)$. Proposition 17 implies that all affine invariants of γ_F (as weight, differential and linear properties, algebraic degree, et al.) can be used as CCZ-invariants for F.

Hence, although studying γ_F is interesting by itself (see [13]), there are also practical reasons for it: because they can be a source of potentially new bent functions when F is AB and because affine invariants of γ_F are CCZ-invariants for F. However, in general it may be a difficult matter to determine γ_F for a given F. Up to now representations of γ_F have been determined only for Gold and inverse power APN functions while determining γ_F for the rest of AB power functions has been an open problem, see [13]. In the present section we find the representation of γ_F for all known power AB functions and for almost all known families of APN polynomials. We also try to determine whether these bent functions (when F is AB) belong to the main known classes of bent functions. The left open cases of APN functions to determine γ_F are the Dobbertin power APN functions and APN polynomials (8-10) from Table 2.6.

4.3.1 Power AB Functions

The representations of functions γ_F were found for Gold and inverse power functions in [13]:

1. Let $F(x) = x^{2^i+1}$, $\gcd(i, n) = 1$, then

$$\gamma_F(a, b) = \mathrm{tr}_n\left(\frac{b}{a^{2^i+1}}\right) \quad \text{with} \frac{1}{0} = 0.$$

2. Let $F(x) = x^{2^n-2}$ then

$$\gamma_F(a, b) = \mathrm{tr}_n\left(\frac{1}{ab}\right) + 1 + \Delta_0(a) + \Delta_0(b) + \Delta_0(a)\Delta_0(b) + \Delta_0(ab + 1)$$

where $\Delta_0(x)$ equals 1 for $x = 0$ and 0 otherwise.

Below we find the function γ_F for Kasami, Welch and Niho power APN functions using the works of Dobbertin [17–19] and by this we solve the open problem mentioned in [13].

Welch Case Let $n = 2m + 1$ and $F(x) = x^d$ with $d = 2^m + 3$. Then according to [18] we have

$$F(x + 1) + F(x) + 1 = q(x^{2^m} + x)$$

for a permutation polynomial

$$q(x) = x^{2^{m+1}+1} + x^3 + x.$$

Then the equation $F(x + 1) + F(x) = b$ is equivalent to the equation

$$q(x^{2^m} + x) = b + 1.$$

Hence,

$$\gamma_F(a, b) = \gamma_F(1, b/a^d) = \begin{cases} \mathrm{tr}_n(q^{-1}(b/a^d + 1)) + 1 & \text{if } a \neq 0, \\ 0 & \text{otherwise.} \end{cases}$$

Niho Case Let $n = 2m + 1$ and $r = m/2$ if m is even and $r = (3m + 1)/2$ if m is odd (i.e. $4r + 1 \equiv 0 \bmod n$). Then the Niho power function over $\mathbb{F}_{2^n}$ is $F(x) = x^d$ with $d = 2^{2r} + 2^r - 1$. The APN property of the Niho function was proven by Dobbertin [17] and we get γ_F using this proof.

We have

$$F(x + 1) + F(x) + 1 = \begin{cases} \frac{1}{q((x^{2^r} + x)^{2^r-1}+1)} & \text{if } x \notin \mathbb{F}_2 \\ 0 & \text{otherwise} \end{cases} \tag{4.8}$$

where

$$q(x) = x^{2^{2r+1}+2^{r+1}+1} + x^{2^{2r+1}+2^{r+1}-1} + x^{2^{2r+1}+1} + x^{2^{2r+1}-1} + x$$

is a permutation. Using (4.8) we get

$$\gamma_F(a,b) = \begin{cases} 1 & \text{if } b = a^d, a \neq 0, \\ \text{tr}_n\left(\left(q^{-1}\left(\frac{a^d}{b+a^d}\right) + 1\right)^{\frac{1}{2^r-1}}\right) + 1 & \text{otherwise.} \end{cases}$$

Kasami Case   Let $F(x) = x^d$ with $d = 2^{2i} - 2^i + 1, \gcd(i,n) = 1, 2 \leq i < n/2$. The APN property of the Kasami function was proven by Dobbertin [19]. According to the proof, we have

$$F(x+1) + F(x) + 1 = 1/q_\alpha(x^{2^i} + x)$$

for a permutation polynomial q_α such that

$$q_\alpha(x) = \left(\sum_{j=1}^{i'} x^{2^{ji}} + \alpha\, \text{tr}_n(x)\right) x^{2^n - 2^i},$$

where $i' \equiv 1/i \mod n$, and, $\alpha = 0$ if i' is odd and $\alpha = 1$ otherwise. Then the equation $F(x+1) + F(x) = b$ is equivalent to the equation

$$\frac{1}{q_\alpha(x^{2^i} + x)} = b + 1,$$

or

$$x^{2^i} + x = q_\alpha^{-1}\left(\frac{1}{b+1}\right).$$

Hence

$$\gamma_F(a,b) = \gamma_F(1,b/a^d) = \begin{cases} \text{tr}_n\left(q_\alpha^{-1}\left(\frac{a^d}{b+a^d}\right)\right) + 1 & \text{if } b \neq a^d, a \neq 0, \\ 1 & \text{if } b = a^d, a \neq 0, \\ 0 & \text{if } a = 0. \end{cases}$$

For $\alpha = 0$, Dobbertin determined the inverse of the permutation q_α:

$$q_0^{-1}(1/x) = R(x) = \sum_{i=1}^{i'} A_i(x) + B_{i'}(x),$$

where

$$A_1(x) = x,$$
$$A_2(x) = x^{2^i+1},$$

$$A_{k+2}(x) = x^{2^{(k+1)i}} A_{k+1}(x) + x^{2^{(k+1)i} - 2^{ik}} A_k(x), \quad k \geq 1,$$

$$B_1(x) = 0,$$

$$B_2(x) = x^{2^i - 1},$$

$$B_{k+2}(x) = x^{2^{(k+1)i}} B_{k+1}(x) + x^{2^{(k+1)i} - 2^{ik}} B_k(x), \quad k \geq 1.$$

Hence for $\alpha = 0$ we have

$$\gamma_F(a,b) = \gamma_F(1, b/a^d) = \begin{cases} \mathrm{tr}_n\left(R(b/a^d + 1)\right) + 1 & \text{if } a \neq 0, \\ 0 & \text{otherwise.} \end{cases}$$

Relation to Completed MM Class and to $PSap$ Many bent functions found in trace representation (listed e.g. in [10] and recalled in Sect. 2.2.1) are in the completed MM class. However, bent functions γ_F corresponding to Kasami (when n is odd), Welch and Niho functions are not in general in the completed MM class. Indeed, Kasami, Welch and Niho functions are CCZ-inequivalent to generalized crooked functions for $n = 7, 9$ (see [21]). In Proposition 18 below, we observe that, for n odd, a function F is CCZ-equivalent to a generalized crooked function if and only if γ_F is in the completed MM class. Hence, for $n = 7, 9$ the functions γ_F related to the Kasami, Welch and Niho functions are not in the completed MM class.

For $n = 7, 9$ the functions γ_F related to the Kasami, Welch and Niho functions have algebraic degree $n - 1$ and, therefore, are not in PS_{ap}.

4.3.2 Quadratic APN and AB Functions

For any quadratic function, for every non-zero vector a,

$$F(x) + F(x + a) = \varphi_F(x, a) + F(0) + F(a)$$

where

$$\varphi_F(x, y) = F(0) + F(x) + F(y) + F(x + y)$$

is bilinear. The set

$$E_a = \{F(x) + F(x + a) : \ x \in \mathbb{F}_2^n\}$$

is then an affine subspace of $\mathbb{F}_2^n$. If F is APN, then E_a has cardinality 2^{n-1} and so is an affine hyperplane. Therefore, there exists a unique non-zero vector $G(a)$ and a unique bit $g(a)$ such that

$$E_a = \{y \in \mathbb{F}_2^n | \ G(a) \cdot y = g(a)\}.$$

Hence

$$\gamma_F(a,b) = G(a) \cdot b + g(a) + 1$$

for every $a \neq 0$ and every b. Defining $G(0) = 0$ and $g(0) = 1$ makes this equality valid for any a and b.

For n odd, the fact that γ_F is in MM class (in the sense that it is affine with respect to b, for every a) is characteristic of generalized crooked functions.

Proposition 18 [7] *A function F is generalized crooked if and only if γ_F is affine with respect to b. If n is odd then F is CCZ-equivalent to a generalized crooked function if and only if γ_F is in the completed MM class of bent functions.*

Hence the open problem of existence of non-quadratic generalized crooked functions is equivalent to the question of existence of non-quadratic APN functions whose γ functions are affine with respect to b.

Below we determine γ function for the known classes of quadratic APN functions.

Cases (1) and (2) of APN Polynomials Below we find the γ functions for the family of APN binomials (1) and (2) of Table 2.6 constructed in [4]. These functions are AB for n odd, and therefore, their γ functions are bent.

Let s, k be positive integers, k odd, such that $\gcd(k, 3) = \gcd(s, 3k) = 1$, and $i = sk \bmod 3$, $t = 3 - i$, $n = 3k$, and α a primitive element of $\mathbb{F}_{2^n}^*$. We consider the function

$$F(x) = x^{2^s+1} + \alpha^{2^k-1} x^{2^{ik}+2^{tk+s}} \tag{4.9}$$

which is APN on $\mathbb{F}_{2^n}$ and which is an AB permutation when n is odd (see Theorem 23 and [4]).

Assume first that $i = 2$. For every $u, v \in \mathbb{F}_{2^n}$, $v \neq 0$, we consider the equation $F(x) + F(x + v) = u$. Following the same steps as in the proof of Theorem 23 we get that this equation has solutions if and only if the equation

$$\Delta_a(x) = d \tag{4.10}$$

has solutions for

$$a = \left(\alpha v^{2^{k+s}+2^s+1-2^k(2^s-1)}\right)^{2^k-1}, \tag{4.11}$$

$$d = (F(v) + u)v^{-(2^s+1)}, \tag{4.12}$$

$$\Delta_a(x) = a\left(x^{2^{2k}} + x^{2^{k+s}}\right) + x^{2^s} + x. \tag{4.13}$$

Denoting $b = a^{2^k}$ and

$$T_1(d) = bd^{2^s} + a^{2^s}d^{2^k},$$

$$T_2(d) = \frac{1}{a(b+1)}\left(bd + ad^{2^k} + abd^{2^{2k}}\right),$$

$$T_3(d) = T_1(d) + a^{2^s}b(T_2(d))^{2^s},$$

$$T_4(d) = T_3(d) + \frac{a^{2^s}(b+1)^{2^s} + (a+1)^{2^s}b}{(b+1)^{2^s}}T_2(d),$$

$$P(a) = \frac{(ab)^{2^s+1} + (ab)^{2^s} + a^{2^s}b + a^{2^s} + ab + b}{(b+1)^{2^s+1}a},$$

$$T_5(d) = (T_4(d))^{2^s} + P(a)^{2^s} T_2(d),$$

$$T_6(d) = \frac{a+1}{ab+a} P(a)^{2^s-1} T_4(d) + T_5(d),$$

we get (again following the proof of Theorem 23) that any solution of (4.10) is also
a solution for

$$P(a)^{2^s}(b+1)^{2^s}\left(x^{2^{2s}} + x^{2^s}\right) = T_6(d), \tag{4.14}$$

where $P(a)^{2^s}(b+1) \neq 0$.
    Denote:

$$f(d) = \mathrm{tr}_n(P(a)^{-2^s}(b+1)^{-2^s} T_6(d)). \tag{4.15}$$

Since the function $T_6(d)$ is linear then either $f \equiv 0$ or it is balanced. We conjecture
that $f(d)$ is balanced for any a satisfying (4.11). Hence, since F is APN, and if the
conjecture is true, then (4.10) has solutions if and only if (4.14) does. That is,

$$\gamma_F(v, u) = \begin{cases} 1 + f\left(\frac{F(v)+u}{v^{2^s+1}}\right) & \text{if } v \neq 0, \\ 0 & \text{othewise.} \end{cases} \tag{4.16}$$

For $i = 1$ we get the function $F(x) = x^{2^s+1} + \alpha^{2^k-1} x^{2^k+2^{2k+s}}$ and

$$F_1(x) = \alpha^{-2^{2k}(2^k-1)} F(x)^{2^{2k}} = \alpha^{-2^{2k}(2^k-1)} x^{2^{2k}(2^s+1)} + x^{2^{k+s}+1},$$

that is,

$$\gamma_F(v, u) = \gamma_{F_1}(v, (u\alpha^{1-2^k})^{2^{2k}}),$$

and F_1 is of the type (4.9) with $i = 2$ and $s' = s + k$.
    Using the proof of Theorem 24 we similarly find the γ functions for the binomials
with n divisible by 4.

Case (5) of APN Polynomials Let n be any positive integer and $c \in \mathbb{F}_{2^n}^*$. Then
according to [5, 6] the function $F(x) = x^3 + c^{-1}\mathrm{tr}_n(c^3 x^9)$ is APN over $\mathbb{F}_{2^n}$ (and AB
for n odd). It is also proven in [5, 6] that for n even F defines two different (up to
CCZ-equivalence) functions: with $c = 1$ and c a primitive element of $\mathbb{F}_{2^n}$, while for
n odd only one function (see Sect. 5.4).
    We have:

$$F(x + a) + F(x) + F(a) = a^2 x + ax^2 + c^{-1}\mathrm{tr}_n(c^3(a^8 x + ax^8))$$
$$= a^3(y + c^{-1}a^{-3}\mathrm{tr}_n(c^3 a^9(y + y^2 + y^4))),$$

where $y = (x/a)^2 + x/a$. Denoting

$$F'_a(x) = x + c^{-1}a^{-3}\text{tr}_n(c^3 a^9 (x + x^2 + x^4))$$

we easily compute

$$F'_a \circ F'_a(x) = x + c^{-1}a^{-3}\text{tr}_n(c^3 a^9 (x + x^2 + x^4))\text{tr}_n(c^{-1}a^{-3}).$$

Let $\text{tr}_n(c^{-1}a^{-3}) = 0$. Then F'_a is involutive, and $F(x+a) + F(x) = b$ has solutions if and only if

$$y = F'_a(a^{-3}(b + F(a)))$$

satisfies $\text{tr}_n(y) = 0$, that is,

$$\gamma_F(a,b) = 1 + \text{tr}_n\left(F'_a\left(a^{-3}(b + F(a))\right)\right) = 1 + \text{tr}_n(a^{-3}b + 1).$$

Let $\text{tr}_n(c^{-1}a^{-3}) = 1$. Denoting

$$d = a^{-3}(b + F(a)) = a^{-3}b + 1 + c^{-1}a^{-3}\,\text{tr}_n(c^3 a^9)$$

we get that $F'_a(y) = d$ if and only if y equals d if

$$\text{tr}_n(c^3 a^9 (d + d^2 + d^4)) = 0$$

and

$$d + c^{-1}a^{-3} \text{ if } \text{tr}_n(c^3 a^9(d + c^{-1}a^{-3} + d^2 + c^{-2}a^{-6} + d^4 + c^{-4}a^{-12})) = 1.$$

Note that

$$\text{tr}_n(c^3 a^9(d + c^{-1}a^{-3} + d^2 + c^{-2}a^{-6} + d^4 + c^{-4}a^{-12}))$$
$$= \text{tr}_n(c^3 a^9 (d + d^2 + d^4)) + \text{tr}_n(c^2 a^6 + ca^3 + c^{-1}a^{-3})$$
$$= \text{tr}_n(c^3 a^9 (d + d^2 + d^4)) + 1.$$

Hence $F'_a(y) = d$ has solutions $y \in \{d, d + c^{-1}a^{-3}\}$ if and only if

$$\text{tr}_n(c^3 a^9 (d + d^2 + d^4)) = 0.$$

Since $\text{tr}_n(c^{-1}a^{-3}) = 1$ then $F'_a(y) = d$ has exactly one solution satisfying $\text{tr}_n(y) = 0$. Therefore, when $\text{tr}_n(c^{-1}a^{-3}) = 1$, we have

$$\gamma_F(a,b) = 1 + \text{tr}_n(c^3 a^9 (d + d^2 + d^4)) = 1 + \text{tr}_n(c^3(a^6 b + a^3 b^2 + a^{-3}b^4))$$

(the rest of the terms vanishing). Hence,

$$\gamma_F(a,b) = 1 + \text{tr}_n(c^{-1}a^{-3})\text{tr}_n(c^3(a^6 b + a^3 b^2 + a^{-3}b^4))$$
$$+ (1 + \text{tr}_n(c^{-1}a^{-3}))\text{tr}_n(a^{-3}b + 1)$$
$$= \text{tr}_n(h(a)b) + \text{tr}_n(c^{-1}a^{-3}),$$

where

$$h(a) = \text{tr}_n(c^{-1}a^{-3})(a^{-3} + c^3 a^6 + c^{2^{n-1}+1}a^{2^{n-1}+1} + c^{3 \cdot 2^{n-2}}a^{-3 \cdot 2^{n-2}}) + a^{-3}.$$

Clearly γ_F has algebraic degree $n - 1$ and therefore it is not in PS_{ap}.

Cases (6) and (7) of APN Polynomials Let n be any positive integer divisible by 3 and $c \in \mathbb{F}_{2^n}^*$, $i \in \{1, 2\}$. Then the function

$$F(x) = x^3 + c^{-1}\mathrm{tr}_n^3(c^3 x^9 + c^6 x^{18})^i$$

is APN over $\mathbb{F}_{2^n}$ and it is AB when n is odd (see [5] and Corollary 32). For n even F defines four different (up to CCZ-equivalence) functions and only two for n odd.

For any $a \in \mathbb{F}_{2^n}^*$ we denote

$$v = c^3 a^9,$$
$$w = c^{-1} a^{-3},$$
$$z = ca^3,$$

and we have

$$F(x + a) + F(x) + F(a) = a^3(y^2 + y) + c^{-1}\mathrm{tr}_n^3(v(y^8 + y) + v^2(y^{16} + y^2))^i$$
$$= a^3 t + c^{-1}\mathrm{tr}_n^3((v + v^{2^{n-2}})t + (v + v^2)(t^2 + t^4))^i$$

where $y = x/a$ and $t = y + y^2$. Denoting

$$F_a'(x) = x + w\,\mathrm{tr}_n^3((v + v^{2^{n-2}})x + (v + v^2)(x^2 + x^4))^i,$$
$$d = a^{-3}(F(a) + b) = a^{-3}b + 1 + w\,\mathrm{tr}_n^3(v + v^2)^i,$$

we see that $\gamma_F(a, b) = 1$, $a \in \mathbb{F}_{2^n}^*$, if and only if the system

$$\begin{cases} F_a'(x) = d \\ \mathrm{tr}_n(x) = 0 \end{cases}$$

has solutions. Clearly, if $F_a'(x) = d$ then for some $u \in \mathbb{F}_8$

$$x = d + wu,$$
$$u = \mathrm{tr}_n^3((v + v^{2^{n-2}})d + (v + v^2)(d^2 + d^4))^i$$
$$+ \mathrm{tr}_n^3\left((v + v^{2^{n-2}})wu + (v + v^2)(w^2 u^2 + w^4 u^4)\right)^i.$$

We denote

$$g_u(a, b) = u + \mathrm{tr}_n^3\left((v + v^{2^{n-2}})d + (v + v^2)(d^2 + d^4)\right)^i$$
$$+ \mathrm{tr}_n^3\left((v + v^{2^{n-2}})wu + (v + v^2)(w^2 u^2 + w^4 u^4)\right)^i$$
$$f_u(a, b) = (1 + \mathrm{tr}_n(d + wu))\Delta_0(g_u(a, b)).$$

We see that $F_a'(x) = d$ has a solution satisfying $\text{tr}_n(x) = 0$ if and only if $f_u(a, b) = 1$ for some $u \in \mathbb{F}_8$.

Note that $F_a'(x)$ is a permutation if $\text{tr}_n^3(w) \in \mathbb{F}_2$ and 2-to-1 otherwise. Indeed, since F_a' is linear it is enough to know the number of solutions of $F_a'(x) = 0$. Then we get (using $\text{tr}_n^3(v^{2^{n-2}}w) = \text{tr}_n^3(w^2)$)

$$x = wu, \quad u \in \mathbb{F}_8,$$

$$u = \left(u\, \text{tr}_n^3(z^2 + w^2) + u^2 \text{tr}_n^3(z + z^4) + u^4 \text{tr}_n^3(w + z^2) \right)^i.$$

Hence the number of solutions of $F_a'(x) = 0$ is equal to the number of solutions of

$$x + \left(x\, \text{tr}_n^3(z^2 + w^2) + x^2\, \text{tr}_n^3(z + z^4) + x^4\, \text{tr}_n^3(w + z^2) \right)^i = 0$$

in $\mathbb{F}_8$. Since all coefficients of that equation are in $\mathbb{F}_8$ then it is easy to check with a computer that it has one solution if $\text{tr}_n^3(w) \in \mathbb{F}_2$ and 2 otherwise.

Let α be a primitive element of $\mathbb{F}_8^*$. Since $F_a'(x) = d$ can have at most 2 solutions then

$$\gamma_F(a, b) = \sum_{u \in \mathbb{F}_8} f_0(a, b) f_u(a, b) + \sum_{0 \le j \le k \le 6} f_{\alpha^j}(a, b) f_{\alpha^k}(a, b)$$

if $a \ne 0$ and equals 0 otherwise.

Cases (3), (4) and (11) of APN Polynomials As proven in [11], the families of APN functions (3), (4) and (11) from Table 2.6 are particular cases of Theorem 1 in [11]. Hence, the functions γ related to these three classes as well as all the classes which can be deduced from this theorem can be treated together:

Let i, j be integers such that $\gcd(n/2, i - j) = 1$, and let $s \ne 0$, $t \ne 0$, u and v be elements of $\mathbb{F}_{2^{n/2}}$ such that the polynomial

$$sX^{2^i + 2^j} + uX^{2^i} + vX^{2^j} + t$$

has no root in $\mathbb{F}_{2^{n/2}}$, then

$$F(x, y) = (x\, y, G(x, y)),$$

where

$$G(x, y) = sx^{2^i + 2^j} + ux^{2^i} y^{2^j} + vx^{2^j} y^{2^i} + ty^{2^i + 2^j},$$

is APN [11]. We shall need the following well-known lemma, whose proof is recalled for self-containment:

Lemma 3 [7] *Let i and j be two integers such that $\gcd(n/2, i - j) = 1$, let k be the inverse of $2^{i-j} - 1$ modulo $2^{n/2} - 1$ and let $a \ne 0, b \ne 0, c \in \mathbb{F}_{2^{n/2}}$. Then the equation $ax^{2^j} + bx^{2^i} = c$ has solutions if and only if $\frac{cb^k}{a^{k2^{i-j}}} = \frac{cb^k}{a^{k+1}}$ has null trace.*

Proof The equation $ax^{2^j} + bx^{2^i} = c$ is equivalent to $b^{2^{-j}} x^{2^{i-j}} + a^{2^{-j}} x = c^{2^{-j}}$. Let d be the element of $\mathbb{F}_{2^{n/2}}$ such that $\frac{b^{2^{-j}}}{d} = (\frac{a^{2^{-j}}}{d})^{2^{i-j}}$, that is, $d = (\frac{a^{2^{i-2j}}}{b^{2^{-j}}})^k$, then we have

$$\frac{b^{2^{-j}} x^{2^{i-j}} + a^{2^{-j}} x}{d} = \left(\frac{a^{2^{-j}}}{d} x \right)^{2^{i-j}} + \frac{a^{2^{-j}}}{d} x = \frac{c^{2^{-j}}}{d}$$

and since $\gcd(n/2, i - j) = 1$, this equation admits solutions if and only if $\frac{c^{2^{-j}}}{d}$ has null trace, that is, $\frac{c}{d^{2^j}}$ has null trace. $\square$

For every non-zero (a, b), we have $\gamma_F(a, b) = 1$ if and only if the equation

$$F(x, y) + F(x + a, y + b) = (c, d)$$

has solutions, that is, if and only if the equation

$$
\begin{aligned}
(bx + ay, &s(a^{2^i} x^{2^j} + x^{2^i} a^{2^j}) + u(a^{2^i} y^{2^j} + x^{2^i} b^{2^j}) \\
&+ v(a^{2^j} y^{2^i} + x^{2^j} b^{2^i}) + t(b^{2^i} y^{2^j} + y^{2^i} b^{2^j})) \\
= (bx + ay, &(sa^{2^i} + vb^{2^i})x^{2^j} + (sa^{2^j} + ub^{2^j})x^{2^i} \\
&+ (ua^{2^i} + tb^{2^i})y^{2^j} + (va^{2^j} + tb^{2^j})y^{2^i}) \\
= (c + ab, &d + G(a, b))
\end{aligned}
$$

has solutions. For $a = 0$ and $b \neq 0$, denoting $c' = \frac{c}{b}$, the first equation is equivalent to $x = c'$ and then the second is equivalent to

$$t(b^{2^i} y^{2^j} + b^{2^j} y^{2^i}) = d + tb^{2^i + 2^j} + vb^{2^i} c'^{2^j} + ub^{2^j} c'^{2^i}.$$

Then

$$\gamma_F(0, b) = \mathrm{tr}_{n/2} \left(\frac{(d + tb^{2^i + 2^j} + vb^{2^i} c'^{2^j} + ub^{2^j} c'^{2^i})(tb^{2^j})^k}{(tb^{2^i})^{k+1}} \right) + 1.$$

For $a \neq 0$, replacing x by ax and denoting $c' = \frac{c}{a} + b$, the first equation is equivalent to $y = bx + c'$ and then the second is equivalent to

$$
\begin{aligned}
(sa^{2^i + 2^j} &+ va^{2^j} b^{2^i} + ua^{2^i} b^{2^j} + tb^{2^i + 2^j})x^{2^j} \\
&+ (sa^{2^i + 2^j} + ua^{2^i} b^{2^j} + va^{2^j} b^{2^i} + tb^{2^i + 2^j})x^{2^i} \\
= d + G(a, b) &+ (ua^{2^i} + tb^{2^i})c'^{2^j} + (va^{2^j} + tb^{2^j})c'^{2^i} \\
= d + sa^{2^i + 2^j} &+ tb^{2^i + 2^j} + (ua^{2^i} + tb^{2^i})\frac{c^{2^j}}{a^{2^j}} + (va^{2^j} + tb^{2^j})\frac{c^{2^i}}{a^{2^i}}.
\end{aligned}
$$

Then

$$\gamma_F(a,b) = \text{tr}_{n/2}\left(\frac{g(a,b)}{h(a,b)}\right) + 1,$$

where

$$g(a,b) = \left(d + sa^{2^i+2^j} + tb^{2^i+2^j} + (ua^{2^i} + tb^{2^i})\frac{c^{2^j}}{a^{2^j}} + (va^{2^j} + tb^{2^j})\frac{c^{2^i}}{a^{2^i}}\right)$$
$$\times (sa^{2^i+2^j} + ua^{2^i}b^{2^j} + va^{2^j}b^{2^i} + tb^{2^i+2^j})^k,$$
$$h(a,b) = (sa^{2^i+2^j} + va^{2^j}b^{2^i} + ua^{2^i}b^{2^j} + tb^{2^i+2^j})^{k+1}.$$

4.4   On Dillon's Class H of Bent Functions

Let throughout this section n be even and $n = 2m$. We are going to analyze all known univariate representations of Niho bent functions for their relation to the completed Maiorana-McFarland class. In particular, we prove that two of the known families of Niho bent functions do not belong to the completed MM class [9]. The latter result gives a positive answer to an open problem whether the class of bent functions introduced by Dillon in his thesis of 1974 differs from the completed class of Mayarana-McFarland functions.

As it is previously mentioned, Niho bent functions (2.5) of case (2) (recalled in Sect. 2.2.1) belong to the completed Maiorana-McFarland class. Opposite to that, below we prove that Niho bent functions (2.4) of case (1) do not belong to the completed MM class. This is done using Proposition 2 that gives a criterion for a function to belong to this class. The relation between Niho bent functions of case (3) and the completed MM class still remains unknown.

Theorem 16 [9] *Niho bent functions*

$$f(t) = \text{tr}_m(x^{2^m+1}) + \text{tr}_n(x^d)$$

over $\mathbb{F}_{2^{2m}}$ *with* $d = (2^m - 1)3 + 1$ *do not belong to the completed MM class for* $m > 3$.

Proof Note that $d = 2(2^m + 2^{m-1} - 1)$ and therefore,

$$f(x) = \text{tr}_m(x^{2^m+1}) + \text{tr}_n(x^{d'})$$

for $d' = 2^m + 2^{m-1} - 1$. Note further that

$$d' = \sum_{i=0, i \neq m-1}^{m} 2^i.$$

The 2-weight of d' is equal to m. Assume f belongs to the completed MM class. Then by Proposition 2 there exists an m-dimensional vector space $V \subset \mathbb{F}_{2^{2m}}$ such that the second-order derivatives of f

$$D_{a,b}f(x) = f(x+a+b) + f(x+a) + f(x+b) + f(x)$$

vanish for any $a, b \in V$. Take $m > 3$, select and fix a pair of nonzero $a, b \in V$.

Obviously, $D_{a,b}f$ equals to the sum of the second-order derivative of $\mathrm{tr}_m(x^{2^m+1})$, which is a constant $\mathrm{tr}_m(a^{2^m}b + ab^{2^m})$, and the second-order derivative of $\mathrm{tr}_n(x^{d'})$. Define a set E consisting of all possible numbers e such that $e = \sum_{i=0, i \neq m-1}^{m} e_i 2^i$ with $e_i \in \{0, 1\}$. The binary expansion of any element $e \in E$ is

$$e = (\underbrace{0 \ldots 0}_{m-1} e_m \, 0 \, e_{m-2} \, e_{m-3} \, \ldots \, e_1 \, e_0).$$

If $(d'_{2m-1} \ldots d'_0)$ denotes the binary expansion of d' then the set E is the set of all e such that $0 \le e \le 2^{2m} - 1$ and $e_i \le d'_i$ for all $i, 0 \le i \le 2m - 1$.

Denote $e' = d' - e \in E$ for any $e \in E$. Then we easily get that

$$(x + a)^d = \sum_{e \in E} a^e t^{e'}$$

and then the second-order derivative of $t^{d'}$ is

$$x^{d'} + (x+a)^{d'} + (x+b)^{d'} + (x+a+b)^{d'}$$
$$= x^{d'} + \sum_{e \in E} a^e x^{e'} + \sum_{e \in E} b^e x^{e'} + \sum_{e \in E} (a+b)^e x^{e'}. \tag{4.17}$$

Fix $\bar{e} = 2^2 + 1$ which belongs to E since $m > 3$, and then

$$\bar{e}' = (\underbrace{0 \ldots 0}_{m-1} 1 \, 0 \, 1 \underbrace{\ldots 1}_{m-4} 0 \, 1 \, 0).$$

It is easy to note that all cyclic shifts modulo $(2^n - 1)$ of the binary expansion of $\bar{e}'$ are distinct. Moreover, $\bar{e}'$ is not equal to a cyclic shift of any other vector $e \in E$. Thus, after applying the trace function to (4.17) and collecting coefficients at $x^{\bar{e}'}$, we obtain this equal to $c = a^{\bar{e}} + b^{\bar{e}} + (a+b)^{\bar{e}}$.

Since $D_{a,b}f$ vanishes then the univariate polynomial representing it should be a zero-polynomial. This in particular, means that $c = a^4b + ab^4 = 0$ and $(a/b)^3 = 1$. Thus, $a = \xi^i b$ for some $i \in \{0, 1, 2\}$ and ξ being a primitive element of $\mathbb{F}_4$. If $|V| = 2^m > 4$ then there exists a pair of nonzero elements $a, b \in V$ that are not related like this and which does not allow $D_{a,b}f$ to vanish. This shows that there cannot exist a vector space of dimension greater than 2 such that all second-order derivatives in the direction of its elements vanish. This contradiction completes the proof. $\square$

Note that the cases with $m = 1, 2$ of the function in Theorem 16 result in quadratic bent functions that all belong to the completed MM class. We also checked with a computer that the case $m = 3$ corresponds to the bent function belonging to the completed MM class as well.

Further we will need the folloing lemma.

Lemma 4 [9] *Take even $m > 2$ and interpret $\frac{1}{3}$ as an inverse of 3 modulo $2^m + 1$. Then the exponent $2d_2 = (2^m - 1)\frac{1}{3} + 2$ has the binary weight m.*

Proof First, note that $1/3$ modulo $2^m + 1$ is equal to $(2^m + 2)/3$. Then

$$2d_2 = \frac{2^n - 1}{3} + \frac{2^m - 1}{3} + 2$$

$$= \sum_{i=0}^{m-1} 2^{2i} + \sum_{i=0}^{m/2-1} 2^{2i} + 2$$

$$= \sum_{i=0}^{m/2-1} 2^{2i+1} + \sum_{i=m/2}^{m-1} 2^{2i} + 2$$

whose binary weight equals m if $m > 2$. $\square$

Theorem 17 [9] *Niho bent functions*

$$f(x) = \mathrm{tr}_m(x^{2^m+1}) + \mathrm{tr}_n(x^d)$$

over $\mathbb{F}_{2^{2m}}$ with $d = \frac{1}{6}(2^m - 1) + 1$ and m even, do not belong to the completed MM class for $m > 2$.

Proof It follows from the proof of Lemma 4 that

$$d = 2 \left(\sum_{i=0}^{m/2-2} 2^{2i+1} + \sum_{i=m/2-1}^{m-2} 2^{2i} + 1 \right).$$

Hence, denoting $d' = d/2$ we get

$$f(x) = \mathrm{tr}_m(x^{2^m+1}) + \mathrm{tr}_n(x^{d'}).$$

Further, the proof goes similarly to the previous theorem.

For even $m \geq 4$, define a set E consisting of all possible numbers e with the following binary expansion

$$(0\,0\,0\,\underbrace{e_{2m-4}\,0\,\ldots\,e_m\,0}_{m/2-1}\,e_{m-2}\,e_{m-3}\,\underbrace{0\,e_{m-5}\,\ldots\,0\,e_1}_{m/2-2}\,e_0)$$

with $e_i \in \{0, 1\}$. Denote $e' = d - e \in E$ for any $e \in E$. Then we have

$$(x + a)^d = \sum_{e \in E} a^e x^{e'}.$$

Take $\bar{e} = 2^{m-2} + 1$ and then, since $m > 3$,

$$\bar{e}' = (0\,0\,0\,\underbrace{1\,0\,\ldots\,1\,0}_{m/2-1}\,0\,1\,\underbrace{0\,1\,\ldots\,0\,1}_{m/2-2}\,0).$$

All cyclic shifts modulo $(2^n - 1)$ of the binary expansion of $\bar{e}'$ are distinct. Moreover, $\bar{e}'$ is not equal to a cyclic shift of any other vector $e \in E$. Thus, after applying the trace function to (4.17) and collecting coefficients at $x^{\bar{e}'}$, we obtain this equal to $c = a^{\bar{e}} + b^{\bar{e}} + (a + b)^{\bar{e}}$.

Assume f belongs to the completed MM class. Then by Proposition 2 there exists an m-dimensional vector space $V \subset \mathbb{F}_{2^{2m}}$ such that $D_{a,b}f$ vanish for any $a, b \in V$. Since $D_{a,b}f$ is equal to the sum of a constant and the second-order derivative of $\mathrm{tr}_n(x^{d'})$, then $c = 0$ for all $a, b \in V$. We have $c = a^{2^{m-2}}b + ab^{2^{m-2}} = 0$ and if $ab \neq 0$ then $(a/b)^{2^{m-2}-1} = 1$. Thus, $a = ub$ for some $u \in \mathbb{F}^*_{2^{m-2}}$. Note that $\gcd(m - 2, 2m) \in \{2, 4\}$ for $m > 4$ and is equal to 2 for $m = 4$. Therefore, since $V \subset \mathbb{F}_{2^{2m}}$, for a fixed nonzero $a \in V$ there exist at most 3 values of $b \in V$ that are related to a in this way if $m = 4$ and at most 15 values if $m > 4$. Hence, the dimension of V is strictly smaller than m. This contradiction completes the proof. $\square$

Note that the remaining case with $m = 2$ in Theorem 17 results in the quadratic bent function that obviously belongs to the completed MM class.

Remark 11 [9] Inserting coefficients into the trace terms of the analyzed bent functions does not change anything in the proofs of Theorems 16 and 17. Thus, the results of Theorems 16 and 17 also apply to all Niho bent functions of case (1).

In [16] Dillon showed that the class H intersects with Maiorana-McFarland class and it has remained an open question whether H is contained in completed MM class. Due to Theorems 16 and 17 we can reply this question. Indeed, the class H differs from $\mathcal{H}$ only by a linear term: take $G(z) = F(z) + \mu z$ in the definitions of H and $\mathcal{H}$ and get $f(x, y) = g(x, y) + \mathrm{tr}_m((\mu + 1)y)$. Hence, the completed classes of H (that is, the class of all bent functions EA-equivalent to those in H) and the completed class of $\mathcal{H}$ coincide. On the other hand, the functions of Theorems 16 and 17 belong to $\mathcal{H}$ and do not belong to the completed MM class.

Corollary 15 [9] *Dillon's class H of bent functions is not contained in the completed MM class, that is, there are functions in H which do not belong to the completed MM class.*

4.5 Bent Functions in Odd Characteristics

We prove below that the non-quadratic cases of p-ary bent functions listed in Table 2.1 do not belong to the completed MM class. We also show that in contrast to the binary case, the completed MM class does not cover all quadratic bent functions even when n is even.

Theorem 18 [8] *Let p be any odd prime, k any positive integer, $n = 4k$ and*

$$d = p^{3k} + p^{2k} - p^k + 1.$$

Then the bent function

$$f(x) = \mathrm{tr}_n(x^d + x^2)$$

over $\mathbb{F}_{p^n}$ does not belong to the completed MM class.

Proof Assume that f is EA-equivalent to a function from completed MM class. Then there exists a $2k$-dimensional vector space $V \subset \mathbb{F}_{p^{4k}}$ such that the second order derivative of f

$$D_{a,b}f(x) = f(x + a + b) - f(x + a) - f(x + b) + f(x) \qquad (4.18)$$

vanishes for any $a, b \in V$. Clearly $D_{a,b}f$ equals the sum of the second order derivative of $\mathrm{tr}_n(x^2)$, which is constant and equals $2\mathrm{tr}_n(ab)$, and the second order derivative of $\mathrm{tr}_n(x^d)$.

Note first that

$$d = p^{3k} + 1 + (p-1)\sum_{i=0}^{k-1} p^{k+i}$$

and therefore it has the p-ary representation

$$d = \Big(\underbrace{0\ldots0}_{k-1}\ 1\ \underbrace{0\ldots0}_{k}\ \underbrace{p-1\ldots p-1}_{k}\ \underbrace{0\ldots0}_{k-1}\ 1\Big).$$

In $(x+a)^d$ all the monomials have the form x^t with

$$t = \Big(\underbrace{0\ldots0}_{k-1}\ t_{3k}\ \underbrace{0\ldots0}_{k}\ t_{2k-1}\ldots t_k\ \underbrace{0\ldots0}_{k-1}\ t_0\Big) \qquad (4.19)$$

where $t_0, t_{3k} \in \{0, 1\}$, $t_k, \ldots, t_{2k-1} \in \{0, 1, \ldots, p-1\}$. In $\mathrm{tr}_n((x+a+b)^d - (x+a)^d - (x+b)^d + x^d)$ the coefficient of the monomial $x^{t'}$ with

$$t' = \Big(\underbrace{0\ldots0}_{2k}\ \underbrace{p-1\ldots p-1}_{k}\ \underbrace{0\ldots0}_{k}\Big)$$

is

$$(a+b)^{p^{3k}+1} - a^{p^{3k}+1} - b^{p^{3k}+1} = a^{p^{3k}}b + ab^{p^{3k}}.$$

Indeed, the numbers $t'p^i \bmod (p^n - 1)$, $1 \le i \le n - 1$, are cyclic shifts of the vector t', and it is easy to note that $t' \ne t'p^i \bmod (p^n - 1)$, $1 \le i \le n - 1$. It is also obvious that, any cyclic shift of any vector t, $t \ne t'$, of the form (4.19) is different from t'.

If $D_{a,b}f \equiv 0$ for some $a, b \ne 0$ then all coefficients of the monomials in $D_{a,b}f$ must equal 0, and, in particular, we get for the coefficient of $x^{t'}$

$$a^{p^{3k}}b + ab^{p^{3k}} = 0. \qquad (4.20)$$

If there exists a $2k$-dimensional vector space V such that $D_{a,b}f \equiv 0$ for all $a, b \in V$ then $D_{a,a}f \equiv 0$ for all $a \in V$, and therefore $2a^{p^{3k}+1} = 0$ by (4.20), a contradiction which shows that V cannot exist. $\qquad\qquad\qquad\qquad\qquad\qquad\qquad\qquad\qquad\qquad\qquad\square$

Theorem 19 [8] *Let p be any odd prime, $k \geq 3$ an odd integer, $n = 2k$, ξ primitive in $\mathbb{F}_{3^n}$ and*

$$d = \frac{3^n - 1}{4} + 3^k + 1.$$

Then the ternary bent function

$$f(x) = \mathrm{tr}_n\left(\xi^{\frac{3^k+1}{4}} x^d\right)$$

over $\mathbb{F}_{3^n}$ does not belong to the completed MM class.

Proof We have

$$d = 3^k + 1 + 2\sum_{i=0}^{k-1} 3^{2i}$$

$$= 3 + 3^k + 2\sum_{i=1}^{k-1} 3^{2i}$$

$$= \Big(\underbrace{0\,2\,...\,0\,2}_{(k-1)/2}\,1\,\underbrace{2\,0\,...\,2\,0}_{(k-3)/2}\,2\,1\,0\Big).$$

Assume there exists an $n/2$-dimensional vector space $V \subset \mathbb{F}_{3^n}$ such that

$$D_{a,b}f(x) = \mathrm{tr}_n\left(\xi^{\frac{3^k+1}{4}}((x+a+b)^d - (x+a)^d - (x+b)^d + x^d)\right)$$

vanishes for any $a, b \in V$. Let k be odd. In $(x+a)^d$ all the monomials have the form x^t with

$$t = \Big(0\,t_{n-2}...\,0\,t_{k+1}\,t_k\,t_{k-1}\,0\,...\,t_4\,0\,t_2\,t_1\,0\Big) \qquad (4.21)$$

where $t_1, t_k \in \{0, 1\}$, $t_{2i} \in \{0, 1, 2\}$, $1 \leq i \leq k - 1$. Among them we choose $t' = \Big(0\,2\,...\,0\,2\,0\,0\Big)$, that is, in t' we take $t_1, t_k = 0$, $t_{2i} = 2$, $1 \leq i \leq k - 1$. It is easy to see that $t' \neq t'p^i \bmod (p^n - 1)$, $1 \leq i \leq n - 1$ and that, any cyclic shift of any vector t, $t \neq t'$, of the form (4.21) is different from t'. Therefore, in $\mathrm{tr}_n(\xi^{\frac{3^k+1}{4}}((x+a+b)^d - (x+a)^d - (x+b)^d + x^d))$ the coefficient of the monomial $x^{t'}$ is

$$\xi^{\frac{3^k+1}{4}}\Big((a+b)^{3^k+3} - a^{3^k+3} - b^{3^k+3}\Big) = \xi^{\frac{3^k+1}{4}}\Big(a^{3^k}b^3 + a^3 b^{3^k}\Big).$$

If $D_{a,b}f \equiv 0$ then all coefficients of the monomials in $D_{a,b}f$ must equal 0, and in particular choosing $a = b$ we get $2\xi^{\frac{3^k+1}{4}}a^{3^k+3} = 0$ for any $a \in V$, a contradiction which shows that V cannot exist, and therefore, f does not belong to the completed MM class. $\square$

Theorem 20 [8] *Let p be any odd prime, n a positive integer, $3 \le k \le n$ odd with* $\gcd(k, n) = 1$, $c \in \mathbb{F}_{3^n}^*$ *and*

$$d = \frac{3^k + 1}{2}$$

Then the ternary bent function

$$f(x) = \text{Tr}_n(cx^d)$$

over $\mathbb{F}_{3^n}$ does not belong to the completed MM class.

Proof Note that

$$d = 3^{k-1} + \ldots + 3^2 + 3 + 2 = \Big(\underbrace{0 \ldots 0}_{n-k} \underbrace{1 \ldots 1}_{k-1} 2 \Big).$$

Assume that f is EA-equivalent to a function from the class MM. Then there exists an $n/2$-dimensional vector space $V \subset \mathbb{F}_{3^n}$ such that

$$D_{a,b}f(x) = \text{tr}_n\Big(c\big((x + a + b)^d - (x + a)^d - (x + b)^d + x^d\big)\Big)$$

vanishes for any $a, b \in V$. In $(x + a)^d$ all the monomials have the form x^t with

$$t = \Big(\underbrace{0 \ldots 0}_{n-k} t_{k-1} \ldots t_1 t_0 \Big) \tag{4.22}$$

where $t_0 \in \{0, 1, 2\}$, $t_1, \ldots, t_{k-1} \in \{0, 1\}$. In $D_{a,b}f$ the coefficient of the monomial $x^{t'}$ with

$$t' = \Big(\underbrace{0 \ldots 0}_{n-k+2} \underbrace{1 \ldots 1}_{k-3} 2 \Big)$$

is

$$c\big((a + b)^{3^{k-1}+3^{k-2}} - a^{3^{k-1}+3^{k-2}} - b^{3^{k-1}+3^{k-2}}\big) = c\big(a^{3^{k-1}}b^{3^{k-2}} + a^{3^{k-2}}b^{3^{k-1}}\big).$$

Indeed, the numbers $t'p^i \bmod (p^n - 1)$, $1 \le i \le n - 1$, are cyclic shifts of the vector t', and it is easy to note that $t' \ne t'p^i \bmod (p^n - 1)$, $1 \le i \le n - 1$. It is also obvious that, any cyclic shift of any vector t, $t \ne t'$, of the form (4.22) is different from t'.

If $D_{a,b}f \equiv 0$ for some $a, b \ne 0$ then all coefficients of the monomials in $D_{a,b}f$ must equal 0, and, in particular, the coefficient $c(a^{3^{k-1}}b^{3^{k-2}} + a^{3^{k-2}}b^{3^{k-1}})$ of $x^{t'}$ is 0. If there exists a $2k$-dimensional vector space V such that $D_{a,b}f \equiv 0$ for all $a, b \in V$ then $D_{a,a}f \equiv 0$ for all $a \in V$, and therefore $2ca^{3^{k-1}+3^{k-2}} = 0$, a contradiction. $\square$

Theorem 21 [8] *Let p be any odd prime, k a positive integer, $n = 2k$ and $c \in \mathbb{F}_{p^n}^*$. If the equation*

$$(cx^2)^{p^k-1} = -1$$

has solutions then the bent function

$$f(x) = \mathrm{tr}_n(cx^2)$$

over $\mathbb{F}_{p^n}$ belongs to the completed MM class. In particular, f belongs to the completed MM class if $c^{p^k-1} = -1$.

Proof Let $u \in \mathbb{F}_{p^{2k}}$. Then for any $a, b \in u\mathbb{F}_{p^k}$ we get $a = ua'$, $b = ub'$ for some $a', b' \in \mathbb{F}_{p^k}$, and

$$D_{a,b}f(x) = 2\mathrm{tr}_n(cab) = 2\mathrm{tr}_k\left(a'b'\mathrm{tr}_n^k(cu^2)\right) = 2\mathrm{tr}_k\left(a'b'(cu^2 + (cu^2)^{p^k})\right).$$

If u is a solution for the equation $(cx^2)^{p^k-1} = -1$ then $cu^2 + (cu^2)^{p^k} = 0$ and therefore $D_{a,b}f$ vanishes for any $a, b \in u\mathbb{F}_{p^k}$. In particular if $c^{p^k-1} = -1$ then $D_{a,b}f$ vanishes for any $a, b \in \mathbb{F}_{p^k}$. $\square$

Remark 12 The theorem above has been generalized recently in [14]. For an even integer n, the monomial bent function $f(x) = \mathrm{tr}_n(ax^{p^j+1})$ over $\mathbb{F}_{p^n}$ with $0 \le j \le n$ and $n/\gcd(j, n)$ odd, belongs to the completed MM class if and only if
- $p = 1 \bmod 4$ and a is a nonsquare in $\mathbb{F}_{p^n}$, or
- $p = 3 \bmod 4$, $n = 2 \bmod 4$, and a is a square in $\mathbb{F}_{p^n}$, or
- $p = 3 \bmod 4$, $n = 0 \bmod 4$, and a is a nonsquare in $\mathbb{F}_{p^n}$.

Theorem 22 [8] *Let p be any odd prime, $k \ge 3$ and t positive integers such that $\gcd(t, 3^k + 1) = 1$. Let also $n = 2k$, $c \in \mathbb{F}_{3^n}^*$ such that $K(c^{3^k+1}) = 0$ and*

$$d = t(3^k - 1).$$

Then the bent function

$$f(x) = \mathrm{tr}_n(cx^d)$$

over $\mathbb{F}_{3^n}$ does not belong to the completed MM class when $t = 3^i + w$ for some nonnegative i and $0 \le w \le 2$.

Proof We just sketch the proof here since it is similar to the ones of Theorems 18–20. We can assume without loss of generality that $0 \le i \le k - 1$.

Let first $w = 0$. Then we can assume that $i = 0$. We have

$$d = 3^k - 1 = 2(3^{k-1} + 3^{k-2} + \ldots + 1) = \Big(\underbrace{0 \ldots 0}_{k}\, \underbrace{2 \ldots 2}_{k}\Big).$$

In $(x + a)^d$ all the monomials are of the type x^t with

$$t = (\, 0 \ldots 0\, t_{k-1} \ldots t_0\,)$$

where $t_0, \ldots, t_{k-1} \in \{0, 1, 2\}$. Let

$$t' = \Big(\underbrace{0 \ldots 0}_{k} \ \underbrace{2 \ldots 2}_{k-2} \ 1 \ 1 \Big).$$

It is easy to note that $t' \neq t' p^i \bmod (p^n - 1)$, $1 \leq i \leq n - 1$ and that any cyclic shift of any vector t, $t \neq t'$, of the form above is different from t'. Hence $x^{t'}$ has the coefficient $4a^4$ in $(x + a)^d$, and therefore it has the coefficient

$$4c((a + b)^4 - a^4 - b^4) = 4c(ab^3 + a^3 b)$$

in $D_{a,b} f$. If $D_{a,a} f \equiv 0$ for some $a \in \mathbb{F}_{3^n}^*$ then $a^4 = 0$, a contradiction.

Let $w = 1$. Then we have

$$d = (3^i + 1)(3^k - 1) = 3^{i+k} + 3^i + 2 \sum_{j=0, j \neq i}^{k-1} 3^j$$

$$= \Big(\underbrace{0 \ldots 0}_{k-i-1} \ 1 \ \underbrace{0 \ldots 0}_{i} \ \underbrace{2 \ldots 2}_{k-i-1} \ 1 \ 2 \underbrace{\ldots 2}_{i} \Big).$$

Denoting

$$t' = \Big(\underbrace{0 \ldots 0}_{k} \ \underbrace{2 \ldots 2}_{k-i-1} \ 0 \ \underbrace{2 \ldots 2}_{i} \Big)$$

we see that the monomial $x^{t'}$ has the coefficient

$$c((a + b)^{3^{k+i}+3^i} - a^{3^{k+i}+3^i} - b^{3^{k+i}+3^i}) = c(a^{3^{k+i}} b^{3^i} + a^{3^i} b^{3^{k+i}})$$

in $D_{a,b} f$. If $D_{a,a} f \equiv 0$ for some $a \in \mathbb{F}_{3^n}^*$ then $a^{3^{k+i}+3^i} = 0$, a contradiction.

Let $w = 2$. Then $1 \leq i \leq k - 1$ and we have

$$d = (3^i + 2)(3^k - 1) = 3^{i+k} + 3^k + 3^i + 1 + 2 \sum_{j=1, j \neq i}^{k-1} 3^j$$

$$= \Big(\underbrace{0 \ldots 0}_{k-i-1} \ 1 \ \underbrace{0 \ldots 0}_{i-1} \ 1 \ \underbrace{2 \ldots 2}_{k-i-1} \ 1 \ 2 \underbrace{\ldots 2}_{i-1} \ 1 \Big).$$

Denoting $t' = \Big(\underbrace{0 \ldots 0}_{k-i-1} \ 1 \ \underbrace{0 \ldots 0}_{i-1} \ 1 \ \underbrace{2 \ldots 2}_{k-i-1} \ 0 \ \underbrace{2 \ldots 2}_{i-1} \ 0 \Big)$ we see that the monomial $x^{t'}$ has the coefficient

$$c((a + b)^{3^i+1} - a^{3^i+1} - b^{3^i+1}) = c(ab^{3^i} + a^{3^i} b)$$

in $D_{a,b} f$. If $D_{a,a} f \equiv 0$ for some $a \in \mathbb{F}_{3^n}^*$ then $a^{3^i+1} = 0$, a contradiction. $\square$

References

1. K. A. Browning, J. F. Dillon, M. T. McQuistan, A. J. Wolfe. An APN Permutation in Dimension Six. *Post-proceedings of the 9-th International Conference on Finite Fields and Their Applications Fq'09, Contemporary Math.*, AMS, v. 518, pp. 33–42, 2010.
2. L. Budaghyan and C. Carlet. CCZ-equivalence of single and multi output Boolean functions. *Post-proceedings of the 9-th International Conference on Finite Fields and Their Applications Fq'09*, Contemporary Math., AMS, v. 518, pp. 43–54, 2010.
3. L. Budaghyan, C. Carlet, A. Pott. New Classes of Almost Bent and Almost Perfect Nonlinear Functions. *IEEE Trans. Inform. Theory*, vol. 52, no. 3, pp. 1141–1152, March 2006.
4. L. Budaghyan, C. Carlet, G. Leander. Two classes of quadratic APN binomials inequivalent to power functions. *IEEE Trans. Inform. Theory*, 54(9), pp. 4218–4229, 2008.
5. L. Budaghyan, C. Carlet, G. Leander. On a construction of quadratic APN functions. *Proceedings of IEEE Information Theory Workshop, ITW'09*, pp. 374–378, Taormina, Sicily, Oct. 2009.
6. L. Budaghyan, C. Carlet, G. Leander. Constructing new APN functions from known ones. *Finite Fields and Their Applications*, v. 15, issue 2, pp. 150–159, April 2009.
7. L. Budaghyan, C. Carlet, T. Helleseth. On bent functions associated to AB functions. Proceedings of *IEEE Information Theory Workshop*, ITW'11, Paraty, Brazil, Oct. 2011.
8. L. Budaghyan, C. Carlet, T. Helleseth, A. Kholosha. Generalized Bent Functions and Their Relation to Maiorana-McFarland Class. *Proceedings of the IEEE International Symposium on Information Theory, ISIT 2012*, Cambridge, MA, USA, 1–6 July 2012.
9. L. Budaghyan, C. Carlet, T. Helleseth, A. Kholosha, S. Mesnager. Further Results on Niho Bent Functions. *IEEE Trans. Inform. Theory*, 58(11), pp. 6979–6985, 2012.
10. C. Carlet. Vectorial Boolean Functions for Cryptography. Chapter of the monography *Boolean Methods and Models*, Yves Crama and Peter Hammer eds, Cambridge University Press, pp. 398–469, 2010.
11. C. Carlet. Relating three nonlinearity parameters of vectorial functions and building APN functions from bent functions. *Designs, Codes and Cryptography*, v. 59(1–3), pp. 89–109, 2011.
12. C. Carlet and S. Mesnager. "On Dillon's class H of bent functions, Niho bent functions and o-polynomials," *J. Combin. Theory Ser. A*, vol. 118, no. 8, pp. 2392–2410, Nov. 2011.
13. C. Carlet, P. Charpin and V. Zinoviev. Codes, bent functions and permutations suitable for DES-like cryptosystems. *Designs, Codes and Cryptography*, 15(2), pp. 125–156, 1998.
14. A. Cesmelioglu, W. Meidl, A. Pott. On the normality of p-ary bent functions. *Pre-proceedings of the International Workshop on Coding and Cryptography WCC 2013*, Bergen, Norway, Apr. 2013.
15. J. F. Dillon. Elementary Hadamard Difference sets. Ph. D. Thesis, Univ. of Maryland, 1974.
16. J. F. Dillon and H. Dobbertin, "New cyclic difference sets with Singer parameters," *Finite Fields Appl.*, vol. 10, no. 3, pp. 342–389, Jul. 2004.
17. H. Dobbertin. Almost perfect nonlinear power functions over $GF(2^n)$: the Niho case. *Inform. and Comput.*, 151, pp. 57–72, 1999.
18. H. Dobbertin. Almost perfect nonlinear power functions over $GF(2^n)$: the Welch case. *IEEE Trans. Inform. Theory*, 45, pp. 1271–1275, 1999.
19. H. Dobbertin. Kasami power functions, permutation polynomials and cyclic difference sets, in: A. Pott, P.V. Kumar, T. Helleseth, D. Jungnickel (Eds.), *Difference Sets, Sequences and their Correlation Properties*, NATO Science Series C, Kluwer, Dordrecht, 1999, pp. 133–158.
20. H. Dobbertin, G. Leander, A. Canteaut, C. Carlet, P. Felke, and P. Gaborit, "Construction of bent functions via Niho power functions," *J. Combin. Theory Ser. A*, vol. 113, no. 5, pp. 779–798, Jul. 2006.
21. Y. Edel and A. Pott. A new almost perfect nonlinear function which is not quadratic. *Advances in Mathematics of Communications* 3, no. 1, pp. 59–81, 2009.

22. G. Leander. Monomial bent functions. *IEEE Transactions on Information Theory*, vol. 52, no. 2, pp. 738–743, 2006.
23. K. Nyberg. Differentially uniform mappings for cryptography. *Advances in Cryptography, EUROCRYPT'93*, Lecture Notes in Computer Science 765, pp. 55–64, 1994.

Chapter 5
New Classes of APN and AB Polynomials

5.1 On the Structure of this Chapter

In this chapter we construct 7 out of 11 known infinite families of quadratic APN polynomials CCZ-inequivalent to power functions, 4 of which are also AB when n is odd [11, 14–16]. In Sect. 5.2 we introduce two infinite classes of quadratic APN functions for n divisible by 3, resp. 4. We prove that for n odd these functions are AB permutations. Since AB permutations taking 0 value at 0 are crooked [1], then the introduced AB binomials are crooked. This binomials and the well-known Gold AB functions are the only known families of crooked functions. We show that, for $n \geq 12$, these functions are EA-inequivalent to power mappings and CCZ-inequivalent to Gold, Kasami, inverse and Dobbertin functions. This implies that for n even they are CCZ-inequivalent to all known APN functions. In particular, for $n = 12, 20, 24$, they are CCZ-inequivalent to any power mappings. These classes of binomials are the firstly found classes of APN functions CCZ-inequivalent to power mappings. Besides, they are the first counterexamples for the conjecture of [21] on nonexistence of quadratic AB functions inequivalent to the Gold maps. Further we discuss the possibility of generalization of the introduced APN binomials for other divisors of n.

In Sect. 5.3 we develop the method for constructing differentially 4-uniform quadratic polynomials introduced by Dillon [23] by proposing its various generalizations. We construct a new infinite class of quadratic APN trinomials and a new potentially infinite class of quadratic APN hexanomials which we conjecture to be CCZ-inequivalent to power functions for $n \geq 6$ and we confirm this conjecture for $n \leq 10$.

We present in Sect. 5.4 a method for constructing new quadratic APN functions from known ones. Applying this method to the Gold power functions we construct an APN function $x^3 + \mathrm{tr}_n(x^9)$ over $\mathbb{F}_2^n$. It is proven that for almost all $n \geq 7$ this function is CCZ-inequivalent to the Gold functions, and in the case $7 \leq n \leq 10$ it is CCZ-inequivalent to any power mapping (and, therefore, to any APN function belonging to one of the families of APN functions known so far). This was the first APN polynomial CCZ-inequivalent to power functions with all coefficients in $\mathbb{F}_2$ and is still the only one which is defined for any n.

© Springer International Publishing Switzerland 2014
L. Budaghyan, *Construction and Analysis of Cryptographic Functions*,
DOI 10.1007/978-3-319-12991-4_5

Section 5.5 is a continuation of the previous section. We give sufficient conditions on linear functions L_1 and L_2 from $\mathbb{F}_{2^n}$ to itself such that the function $L_1(x^3)+L_2(x^9)$ is APN over $\mathbb{F}_{2^n}$. We show that this can lead to many new cases of APN functions. In particular, we get two families of APN functions $x^3 + a^{-1}\text{tr}_n^3(a^3x^9 + a^6x^{18})$ and $x^3 + a^{-1}\text{tr}_n^3(a^6x^{18} + a^{12}x^{36})$ over $\mathbb{F}_{2^n}$ for any n divisible by 3 and $a \in \mathbb{F}_{2^n}^*$. We prove that for $n = 9$, these families are pairwise different and differ from all previously known families of APN functions, up to the most general equivalence notion, the CCZ-equivalence. We also investigate further sufficient conditions under which the conditions on the linear functions L_1 and L_2 are satisfied.

5.2   Classes of APN Binomials Over $\mathbb{F}_{2^{3k}}$ and $\mathbb{F}_{2^{4k}}$

Let s, k, p be positive integers such that $\gcd(k, p) = \gcd(s, pk) = 1$, and $i = sk \bmod p, t = p - i, n = pk$. In this section we prove that the function

$$F(x) = x^{2^s+1} + wx^{2^{ik}+2^{tk+s}} \tag{5.1}$$

is APN on $\mathbb{F}_{2^n}$ when $p = 3, 4$ (under some conditions on the element $w \in \mathbb{F}_{2^n}^*$) and we also show that, most probably, other values of parameter p do not define classes of APN functions.

APN Binomials Over $\mathbb{F}_{2^{3k}}$ The following theorem introduces a large class of quadratic binomial APN functions for n divisible by 3.

Theorem 23   [14] *Let s and k be positive integers with $\gcd(s, 3k) = 1$, and $t \in \{1, 2\}$, $i = 3 - t$, $n = 3k$. Furthermore let*

$$d = 2^{ik} + 2^{tk+s} - (2^s + 1),$$

$$g_1 = \gcd(2^{3k} - 1, d/(2^k - 1)),$$

$$g_2 = \gcd(2^k - 1, d/(2^k - 1)),$$

and let $w \in \mathbb{F}_{2^n}^*$ have the order $2^{2k} + 2^k + 1$ (i.e. $w = \alpha^{2^k-1}$ for some primitive element α of $\mathbb{F}_{2^n}^*$). If $g_1 \neq g_2$ then function (5.1) is almost perfect nonlinear on $\mathbb{F}_{2^n}$ (and is almost bent when k is odd).

Proof   To prove the theorem it is enough to consider only the case $i = 2$. Indeed, for $i = 1$ we get the function

$$F(x) = x^{2^s+1} + wx^{2^k+2^{2k+s}}$$

which is EA-equivalent to

$$F'(x) = w^{-2^{2k}} F(x)^{2^{2k}} = w^{-2^{2k}} x^{2^{2k}(2^s+1)} + x^{2^{k+s}+1}.$$

The coefficient $w^{-2^{2k}}$ has the order $2^{2k}+2^k+1$ if and only if w does, and it is easy to note that F' is of the type (5.1) with $i=2$ and $s'=s+k$, $d'=-2^{2k}d \bmod (2^{3k}-1)$. So F is APN if and only if F' is APN.

Thus, without loss of generality we assume $i=2$. We have to show that for every $u,v \in \mathbb{F}_{2^n}$, $v \neq 0$, the equation

$$F(x)+F(x+v)=u$$

has at most 2 solutions. We have

$$
\begin{aligned}
F(x)+F(x+v) &= \alpha^{2^k-1}\left(x^{2^{2k}+2^{k+s}}+(x+v)^{2^{2k}+2^{k+s}}\right) \\
&\quad + x^{2^s+1}+(x+v)^{2^s+1} \\
&= \alpha^{2^k-1}v^{2^{2k}+2^{k+s}}\left(\left(\frac{x}{v}\right)^{2^{2k}}+\left(\frac{x}{v}\right)^{2^{k+s}}\right) \\
&\quad + v^{2^s+1}\left(\left(\frac{x}{v}\right)^{2^s}+\left(\frac{x}{v}\right)\right)+\alpha^{2^k-1}v^{2^{2k}+2^{k+s}}+v^{2^s+1}.
\end{aligned}
$$

As this is a linear equation in x it is sufficient to study the kernel of the corresponding linear mapping. Note furthermore that

$$v^{2^{2k}+2^{k+s}-(2^s+1)}=v^{(2^k-1)(2^{k+s}+2^s+1-2^k(2^s-1))}.$$

To simplify notation we define

$$a=\left(\alpha v^{2^{k+s}+2^s+1-2^k(2^s-1)}\right)^{2^k-1}.$$

After replacing x by vx and dividing by v^{2^s+1}, we finally see that the equation $F(x)+F(x+v)=u$ admits 0 or 2 solutions for every $v \in \mathbb{F}_{2^n}^*$ if and only if, denoting

$$\Delta_a(x)=a\left(x^{2^{2k}}+x^{2^{k+s}}\right)+x^{2^s}+x,$$

the equation $\Delta_a(x)=0$ has at most two zeros or, equivalently, that the only solutions are $x=0$ and $x=1$.

The following step can be seen as a very basic application of the multivariate method introduced by Dobbertin [26]. If we denote $y=x^{2^k}$, $z=y^{2^k}$ and $b=a^{2^k}$, $c=b^{2^k}$ the equation $\Delta_a(x)=0$ can be rewritten as

$$a(z+y^{2^s})+(x^{2^s}+x)=0.$$

By definition, a is always a (2^k-1)-th power and thus $abc=1$. Besides, $a \notin \mathbb{F}_2$ (as it is confirmed further). Considering also the conjugated equations we derive the following system of equations

$$
\begin{aligned}
f_1 &= \Delta_a(x) = a(z+y^{2^s})+x^{2^s}+x &= 0 \\
f_2 &= f_1^{2^k} &= b(x+z^{2^s})+y^{2^s}+y &= 0 \\
f_3 &= f_1^{2^{2k}} &= \tfrac{1}{ab}(y+x^{2^s})+z^{2^s}+z &= 0.
\end{aligned}
$$

The aim now is eliminating y and z from these equations and finally getting an equation in x only. First we compute

$$R_1 = b(f_1)^{2^s} + a^{2^s} f_2$$
$$= a^{2^s} b y^{2^{2s}} + a^{2^s} y^{2^s} + a^{2^s} y + b x^{2^{2s}} + b x^{2^s} + a^{2^s} b x$$

and

$$R_2 = \frac{1}{a(b+1)} (bf_1 + af_2 + abf_3)$$
$$= y^{2^s} + \frac{a+1}{ab+a} y + \frac{1}{a} x^{2^s} + \frac{ab+b}{ab+a} x$$

to eliminate z. To eliminate $y^{2^{2s}}$ we compute

$$R_3 = R_1 + a^{2^s} b(R_2)^{2^s} = \frac{a^{2^s}(b+1)^{2^s} + (a+1)^{2^s} b}{(b+1)^{2^s}} y^{2^s}$$
$$+ a^{2^s} y + \frac{a^{2^s} b^{2^s+1} + b}{b^{2^s} + 1} x^{2^s} + a^{2^s} bx.$$

Using equations R_2 and R_3 we can eliminate y^{2^s} by computing

$$R_4 = R_3 + \frac{a^{2^s}(b+1)^{2^s} + (a+1)^{2^s} b}{(b+1)^{2^s}} R_2$$
$$= P(a)(y + (b+1)x^{2^s} + bx),$$

where

$$P(a) = \frac{(ab)^{2^s+1} + (ab)^{2^s} + a^{2^s} b + a^{2^s} + ab + b}{(b+1)^{2^s+1} a}.$$

Computing

$$R_5 = (R_4)^{2^s} + P(a)^{2^s} R_2 = P(a)^{2^s}$$
$$\times \left(\frac{a+1}{ab+a} y + (b^{2^s} + 1)x^{2^{2s}} + \frac{ab^{2^s} + 1}{a} x^{2^s} + \frac{ab+b}{ab+a} x \right)$$

we finally get our desired equation by

$$R_6 = \frac{a+1}{ab+a} P(a)^{2^s-1} R_4 + R_5$$
$$= P(a)^{2^s} (b+1)^{2^s} \left(x^{2^{2s}} + x^{2^s} \right).$$

Obviously if x is a solution of $\Delta_a(x) = 0$ then $R_6(x) = 0$. For $P(a)^{2^s}(b+1) \neq 0$ this is equivalent to $x = 0, 1$. Thus to prove the theorem, it is sufficient to show that $P(a)$ does not vanish for elements a fulfilling the equation

$$a = \left(\alpha v^{2^k + 2^s + 1} \right)^{2^k - 1} \tag{5.2}$$

Note that, if a satisfies (5.2), then a is not a $(2^k + 2^s + 1)$-th power, since α^{2^k-1} is not: $g_2 = \gcd(2^k - 1, 2^k + 2^s + 1)$ is by hypothesis a strict divisor of $g_1 = \gcd(2^n - 1, 2^k + 2^s + 1)$ and α being a primitive element, it cannot be a (g_1/g_2)-th power.

Consequently, it is sufficient to show, that if $P(a) = 0$ then a is a $(2^k + 2^s + 1)$-th power. For $a \notin \mathbb{F}_2$ the equation $P(a) = 0$ is equivalent to

$$a = \left(\frac{a+1}{c+1} \right)^{2^s + 1} c^{2^s + 1} \left(\frac{b+1}{a+1} \right) a = \left(\frac{a+1}{c+1} c \right)^{2^k + 2^s + 1},$$

as can be easily seen by dividing this equality by a, simplifying it by $(a + 1)$, and then expanding it, using that $c = 1/ab$. Note that the right hand side is always a $(2^k + 2^s + 1)$-th power. This proves the theorem.   □

From Theorem 23 we get the following corollary as a special case.

Corollary 16 [14] *Let s and k be positive integers such that* $\gcd(k, 3) = \gcd(s, 3k) = 1$, *and $i = sk$ mod 3, $t = 3 - i$, $n = 3k$, and $w \in \mathbb{F}_{2^n}^*$ have the order $2^{2k} + 2^k + 1$. Then the function*

$$F(x) = x^{2^s + 1} + wx^{2^{ik} + 2^{tk+s}}$$

is APN on $\mathbb{F}_{2^n}$ (and is AB when n is odd).

Proof We only have to verify that in this case the greatest common divisors

$$g_1 = \gcd(2^n - 1, 2^{k+s} + 2^s + 1 - 2^k(2^s - 1)(i - 1))$$

$$g_2 = \gcd(2^k - 1, 2^{k+s} + 2^s + 1 - 2^k(2^s - 1)(i - 1))$$

are not equal. Obviously g_2 is always coprime with 7 and it can be easily checked that g_1 is always divisible by 7. Indeed, for instance, if k mod $3 = s$ mod $3 = 1$ then $i = 1$ and $k = 3k' + 1, s = 3s' + 1$ for some k', s', and we get

$$g_1 = 2^{k+s} + 2^s + 1 = 4(2^{3(k'+s')} - 1) + 2(2^{3s'} - 1) + 7.$$

                                    □

The next proposition shows that the functions from Corollary 16 are permutations if k is odd. Moreover computer investigations show that most probably, if k is odd their inverses have algebraic degree $(3k + 1)/2$.

Proposition 19 [14] *The APN functions of Corollary 16 are bijective if and only if k is odd.*

Sketch of Proof If k is even then, since $\gcd(s, 3k) = 1$, s must be odd and therefore $2^s + 1$ is divisible by 3 as well as $2^{ik} + 2^{ik+s} = 2^{ik}(1 + 2^{(t-i)k+s})$. We have $F(x) = F(\gamma x)$ for every $\gamma \in \mathbb{F}_4^*$.

To prove that F is bijective when k is odd, we use the same steps as in the proof of Theorem 23. Assume $i = 1$ (the proof for the case $i = 2$ is similar). We have to show that the equation $F(x) + F(x + v) = 0$ does not have a non zero solution v for any x. Doing the same computations as in the proof of Theorem 23 we have this time to look at the following system of equations

$$f_1 = a(z + y^{2^s} + 1) + x^{2^s} + x + 1 = 0$$

$$f_2 = b(x + z^{2^s} + 1) + y^{2^s} + y + 1 = 0$$

$$f_3 = \frac{1}{ab}(y + x^{2^s} + 1) + z^{2^s} + z + 1 = 0.$$

Now, doing the same elimination of y and z as before, we end up with

$$P(a)^{2^s}(x^{2^s} + x + 1) = 0,$$

where P is as in the proof of Theorem 23. By taking the power 2^s of $x^{2^s} + x + 1 = 0$ and substituting $x^{2^s} = x + 1$ we get $x^{2^{2s}} = x$ which is equivalent to $x \in \mathbb{F}_{2^j}$ where $j = \gcd(2s, 3k)$. If k is odd then $j = 1$ and the only possible solutions could be 0 or 1 but they obviously do not satisfy $x^{2^s} + x + 1 = 0$. □

Remark 13 1. The APN binomials of Corollary 16 have been generalized to APN trinomials in [5]. Let $v \in \mathbb{F}_{2^k}$ and the function F be as in Corollary 16 with $i = 1$, $w = \alpha^{2^k-1}$ (where α is primitive in $\mathbb{F}_{2^{3k}}$), then the function $F(x) + \alpha^{2^k} v x^{2^{k+s}+2^s}$ is also APN (it is a particular case of functions (8–10) in Table 2.6). In [7] these trinomials were extended to the family of quadrinomials corresponding to cases (8–10) in Table 2.6 of known APN polynomials. Note that Theorem 23 covers a larger class of APN functions than the one of Corollary 16 as can be seen by checking the conditions on the greatest common divisors for small values of k and s, that is, APN quadrinomials (8–10) of Table 2.6 do not cover all the functions of Theorem 23.

2. Since for n odd the functions of Corollary 16 are AB then it completely determines their Walsh spectrum. It is shown in [6] that for n even these functions have the same Walsh spectrum as Gold functions, that is, $\{0, \pm 2^{n/2}, \pm 2^{(n+2)/2}\}$. In [4] it is proven that the quadrinomials (8–10) have the same property.

3. It is further proven in [8] that when relaxing conditions of $\gcd(s, 3k) = 1$ to $\gcd(s, 3k) = r$ with k/r odd, the binomials of Corollary 16 are still permutations whose derivatives are 2^r-to-1 mappings. It is left as a problem to prove that the APN quadrinomials (8–10) have the same properties as the binomials when relaxing the conditions.

4. Since every quadratic AB permutation taking 0 value at 0 is crooked then the AB binomials of Theorem 23 (and Corollary 16) are crooked. This binomials and the well-known Gold AB functions are the only known families of crooked functions.

APN Binomials Over $\mathbb{F}_{2^{4k}}$ The following theorem presents another class of APN binomials defined for n divisible by 4.

Theorem 24 [14] *Let s and k be positive integers such that $s \le 4k - 1$, $\gcd(k, 2) = \gcd(s, 2k) = 1$, and $i = sk \bmod 4$, $t = 4 - i$, $n = 4k$. If $w \in \mathbb{F}_{2^n}^*$ has the order $2^{3k} + 2^{2k} + 2^k + 1$ (i.e. $w = \alpha^{2^k-1}$ for some primitive element α of $\mathbb{F}_{2^n}^*$) then the function*

$$F(x) = x^{2^s+1} + wx^{2^{ik}+2^{tk+s}}$$

is APN on $\mathbb{F}_{2^n}$.

Proof Repeating the first steps of Theorem 23 we see that F is APN if and only if, for every $u \in \mathbb{F}_{2^n}^*$ the equation $\Delta_a(x) = 0$ with

$$a = \alpha^{2^k-1} u^{2^{ik}+2^{tk+s}-2^s-1},$$

$$\Delta_a(x) = a\left(x^{2^{ik}} + x^{2^{tk+s}}\right) + x^{2^s} + x,$$

has the only solutions 0 and 1.

From now on we consider the cases $i = 1$ and $i = 3$ separately.

Case 1 ($i = 3, t = 1$) Applying Dobbertin's multivariate method we denote $y = x^{2^k}, z = y^{2^k}, t = z^{2^k}, b = a^{2^k}, c = b^{2^k}, d = c^{2^k}$, and then rewrite the equation $\Delta_a(x) = 0$ as

$$a(t + y^{2^s}) + x^{2^s} + x = 0.$$

Since

$$2^{ik} + 2^{tk+s} - 2^s - 1 \ = 2^{3k} + 2^{k+s} - 2^s - 1$$
$$= (2^k - 1)(2^{2k} + 2^k + 2^s + 1)$$

then the element a is always the $(2^k - 1)$-th power and thus

$$abcd = a^{1+2^k+2^{2k}+2^{3k}} = 1.$$

Considering also the conjugated equations obtained by raising the equation $\Delta_a(x) = 0$ at the powers 2^k, 2^{2k} and 2^{3k} we derive the following system of equations

$$
\begin{aligned}
f_1 &= \Delta_a(x) &&= a(t + y^{2^s}) + x^{2^s} + x &&= 0 \\
f_2 &= f_1^{2^k} &&= b(x + z^{2^s}) + y^{2^s} + y &&= 0 \\
f_3 &= f_2^{2^k} &&= c(y + t^{2^s}) + z^{2^s} + z &&= 0 \\
f_4 &= abcf_3^{2^k} &&= z + x^{2^s} + abc(t^{2^s} + t) &&= 0.
\end{aligned}
$$

Now we eliminate y, z and t from these equations to get an equation in x only. We consider

$$R_1 = bcf_1 + abcf_2 + abf_3 + f_4$$
$$= ab(bc + 1)z^{2^s} + (ab + 1)z + (bc + 1)x^{2^s}$$
$$+ bc(ab + 1)x$$

and

$$R_2 = cf_1^{2^s} + a^{2^s}c(f_2^{2^s} + f_2) + a^{2^s}f_3$$
$$= a^{2^s}b^{2^s}cz^{2^{2s}} + a^{2^s}(bc + 1)z^{2^s} + a^{2^s}z$$
$$+ cx^{2^{2s}} + c(ab + 1)^{2^s}x^{2^s} + a^{2^s}bcx$$

to eliminate t and y. To eliminate $z^{2^{2s}}$ we compute

$$R_3 = cR_1^{2^s} + (bc + 1)^{2^s}R_2$$
$$= (c(ab + 1)^{2^s} + a^{2^s}(bc + 1)^{2^s+1})z^{2^s}$$
$$+ a^{2^s}(bc + 1)^{2^s}z + c(ab + 1)^{2^s}x^{2^s}$$
$$+ a^{2^s}bc(bc + 1)^{2^s}x.$$

Using equations R_1 and R_3 we can eliminate z^{2^s}:

$$R_4 = ab(bc + 1)R_3 + (c(ab + 1)^{2^s} + a^{2^s}(bc + 1)^{2^s+1})R_1$$
$$= P(a)(z + (bc + 1)x^{2^s} + bcx),$$

where

$$P(a) = c(ab + 1)^{2^s+1} + a^{2^s}(bc + 1)^{2^s+1}.$$

Below we shall show that $P(a) \neq 0$, thus we can denote

$$R_5 = \frac{R_4}{P(a)} = z + (bc + 1)x^{2^s} + bcx.$$

Computing

$$R_6 = R_1 + ab(bc + 1)R_5^{2^s}$$
$$= (ab + 1)z + ab(bc + 1)^{2^s+1}x^{2^{2s}}$$
$$+ (ab^{2^s+1}c^{2^s} + 1)(bc + 1)x^{2^s} + bc(ab + 1)x$$

we finally get

$$R_7 = (ab + 1)R_5 + R_6$$
$$= ab(bc + 1)^{2^s+1}\left(x^{2^{2s}} + x^{2^s}\right).$$

If x is a solution of $\Delta_a(x) = 0$ then $R_7(x) = 0$, which is equivalent to $x = 0, 1$ when $P(a) \neq 0$ and $bc + 1 \neq 0$. Thus, it is sufficient to show that $P(a)$ and $bc + 1$ do not vanish when

$$a = \alpha^{2^k - 1} u^{2^{3k} + 2^{k+s} - 2^s - 1}. \tag{5.3}$$

Suppose $bc = 1$, that is, $a^{2^{2k} + 2^k} = 1$ or equivalently $a^{2^k + 1} = 1$. We have

$$a^{2^k + 1} = \left(\alpha u^{2^k + 2^s} \right)^{2^{2k} - 1}$$

because

$$(2^{3k} + 2^{k+s} - 2^s - 1)(2^k + 1) \equiv (2^{2k} - 1)(2^k + 2^s) \bmod (2^{4k} - 1).$$

Since $a^{2^k + 1} = 1$ then $\alpha u^{2^k + 2^s}$ should be the $(2^{2k} + 1)$-th power of an element of the field. We have $2^k + 2^s = 2^s(2^{k-s} + 1) = 2^s(2^{2p} + 1)$ with some p odd. Indeed, $ks \bmod 4 = 3$, then $k \bmod 4 \neq s \bmod 4$ for odd k, s, and $k - s = 2p$ for some p odd.

The numbers $2^{2p} + 1$ and $2^{2k} + 1$ are divisible by 5 because p, k are odd. We get that $u^{2^k + 2^s}$ is the fifth power of an element of the field and $\alpha u^{2^k + 2^s}$ is not (since α is a primitive element). Therefore $\alpha u^{2^k + 2^s}$ is not the $(2^{2k} + 1)$-th power of an element of the field. A contradiction.

Let $c(ab + 1)^{2^s + 1} + a^{2^s}(bc + 1)^{2^s + 1} = 0$. Since $bc + 1 \neq 0$ then $ab + 1 \neq 0$ and we get

$$\frac{c}{a^{2^s}} = \left(\frac{bc + 1}{ab + 1} \right)^{2^s + 1}.$$

Note that since n is even and s is odd then $2^n - 1$ and $2^s + 1$ are divisible by 3. Therefore c/a^{2^s} is the third power of an element of the field. We have

$$c/a^{2^s} = a^{2^{2k} - 2^s} = a^{2^s(2^{2k-s} - 1)}$$

and $2^{3k} + 2^{k+s} - 2^s - 1 = 2^s(2^{3k-s} - 1) + (2^{k+s} - 1)$. The numbers $2^{3k-s} - 1$ and $2^{k+s} - 1$ are divisible by 3 since $3k - s$ and $k + s$ are even. On the other hand $2^k - 1$ and $2^{2k-s} - 1$ are not divisible by 3 since k and $2k - s$ are odd. We get

$$a^{2^s(2^{2k-s} - 1)} = \alpha^{2^s(2^{2k-s} - 1)(2^k - 1)} u^{2^s(2^{2k-s} - 1)(2^{3k} + 2^{k+s} - 2^s - 1)}.$$

Obviously c/a^{2^s} is not the third power of an element of the field and therefore it is not a $(2^s + 1)$-th power. A contradiction.

Case 2 $(i = 1, t = 3)$ It is obvious that the function $F(x) = x^{2^s + 1} + wx^{2^k + 2^{3k+s}}$ is EA-equivalent to $F'(x) = w^{-2^{3k}} x^{2^{3k}(2^s + 1)} + x^{2^{2k+s} + 1}$. The coefficient $w^{-2^{3k}}$ has the order $2^{3k} + 2^{2k} + 2^k + 1$ if and only if w does. Since $sk \bmod 4 = 1$ then $s \bmod 4 = k \bmod 4$ and $(2k + s)k \bmod 4 = 3$. So it follows from the first case that F' is APN, and therefore so is F. $\square$

Remark 14 1. Note that it follows from the proofs of Theorems 23 and 24 that the element w of the function of Corollary 16 and Theorem 24 can be of a more general form, that is, if $n = pk$, $p = 3, 4$, then w has the order h divisible by $\frac{2^n-1}{2^k-1}$ such that $\gcd(\frac{2^n-1}{h}, 2^p - 1) = 1$ (in other words $w = \alpha^e$, where α is some primitive element of $\mathbb{F}_{2^n}^*$, e is a multiple of $2^k - 1$ and co-prime with $2^p - 1$). However, this can hardly give new (up to equivalence) cases of functions.

2. In [2] a common proof of APN properties for the classes of Corollary 16 and Theorem 24 is found. However, this proof is based on the same ideas as the proofs of Corollary 16 and Theorem 24: it does not give better understanding of APN binomials neither leads to more general classes which would include the functions of Corollary 16 and Theorem 24. Moreover, as observed before, not all functions of Theorem 23 are covered by these description.

3. It is shown in [6] that the binomials of Theorem 24 have the same Walsh spectrum as Gold functions, that is, $\{0, \pm 2^{n/2}, \pm 2^{(n+2)/2}\}$.

Further APN Binomials? In [28] the first two quadratic APN functions CCZ-inequivalent to power mappings were introduced. The first one is the binomial

$$x^3 + wx^{36} \tag{5.4}$$

over $\mathbb{F}_{2^{10}}$, where w has the order 3 or 93; the second is $x^3 + wx^{528}$ over $\mathbb{F}_{2^{12}}$, where w has the order 273 or 585 (which gives only one function up to equivalence). The second one is classified into families of APN binomilas by Theorems 23 and 24 while the first one still stays as an isolated example. Below we discuss the existence of further examples of APN binomials different from the ones of Theorems 23 and 24 and the sporadic example (5.4). For simplicity we consider only the case $s = 1$.

Proposition 20 [14] *Let k, p be positive integers and $n = pk$, $\gcd(k, p) = 1$, $i = k \bmod p$, $m = p - i$, α a primitive element of $\mathbb{F}_{2^n}$. Then, for the function*

$$F(x) = x^3 + \alpha^{2^k-1} x^{2^{ik}+2^{ik+1}}$$

over $\mathbb{F}_{2^n}$, the following holds:

1) for $p = 2$ and any $k \geq 2$, the function F is not APN;
2) for $k = 1$ and any p, the function F is EA-equivalent to x^3;
3) for $2 \leq k \leq 30$ and $5 \leq p \leq 30$, the function F is not APN.

Proof For $p = 2$, we have $i = m = 1$ and $F(x) = x^3 + \alpha^{2^k-1} x^{2^k+2^{k+1}} = L(x^3)$ where $L(x) = x + \alpha^{2^k-1} x^{2^k}$ is 2^k-to-1. For any nonzero a, the function $L(x^3) + L((x + a)^3) = L(x^3 + (x + a)^3)$ cannot be 2-to-1 when $k > 1$ (because L is 2^k-to-1). Therefore, F is not APN.

Let $k = 1$ and p be any integer, then $n = p$ and $F(x) = x^3 + \alpha x^{2+2^n} = (1+\alpha)x^3$. The third claim was confirmed with a computer. □

Note that we also tried to find quadratic APN binomials of a more general form, however we could not find any new APN functions this way. The next fact summarizes these (negative) results.

Fact 1 [14] *Let k, $p < 12$, $n = pk$ and α be a primitive element of $\mathbb{F}_{2^n}$. Furthermore, let $0 < i < k$, $0 < j < p$ and d be a divisor of $2^n - 1$. Then the function*

$$F(x) = x^3 + \alpha^d x^{2^{pi}+2^{kj}}$$

is APN over $\mathbb{F}_{2^n}$ if and only if it is equivalent to known cases, i.e. to the Gold cases, to the sporadic case (5.4) or to the classes presented in Theorems 23 and 24.   □

CCZ-Inequivalence to Known Power APN Functions We prove below that the new APN functions introduced in Corollary 16 and Theorem 24 are not CCZ-equivalent to the Gold, Kasami, inverse and Dobbertin functions.

Without loss of generality a Gold function $F(x) = x^{2^s+1}$ and a Kasami function $K(x) = x^{4^r-2^r+1}$ can be considered under conditions $1 \leq s < \frac{n}{2}$, $2 \leq r < \frac{n}{2}$, since this exhausts all different cases (under EA-equivalence). Besides, we can consider only the case $i = 1$ for our binomials as it follows from the proofs of Theorems 23 and 24.

We first prove the EA-inequivalence between the APN binomials and all power functions.

Theorem 25 [14] *Let n be a positive integer and let s, j, q be three nonzero elements of $\mathbb{Z}/n\mathbb{Z}$ such that $q \neq \pm s$. If one of the following conditions holds*

1. $j \notin \{\pm s, \pm q, 2s, s \pm q\}$,
2. $j \notin \{\pm s, \pm q, \pm s - q, -2q\}$,
3. $j \notin \{s, -q, 2s - q, s - 2q, s \pm q, 2s\}$,
4. $j \notin \{s, -q, 2s - q, s - 2q, \pm s - q, -2q\}$,

then the function $F(x) = x^{2^s+1} + ax^{2^j(2^q+1)}$ with $a \in \mathbb{F}_{2^n}^*$ is EA-inequivalent to power functions on $\mathbb{F}_{2^n}$.

Proof Suppose the function F is EA-equivalent to a power function. Since F is quadratic and EA-transformation does not change the algebraic degree of a function then F is EA-equivalent to x^{2^r+1} for some nonzero $r \in \mathbb{Z}/n\mathbb{Z}$ (see [13]). Therefore, there exist affine permutations L_1, L_2 and an affine function L' such that

$$L_1 \circ F = (L_2)^{2^r+1} + L'.$$

Expressing $L_1(x)$, $L_2(x)$ and $L'(x)$ as sums of linearized polynomials and constants and reducing the resulting exponents modulo $2^n - 1$ leads to an equation whose degree is at most $2^{n-1} + 2^{n-2}$ (since the 2-weights of the exponents are at most 2) and which has 2^n solutions. Hence the equation must be an identity.

Since the functions are quadratic, we can assume without loss of generality that L_1 and L_2 are linear:

$$L_1(x) = \sum_{m \in \mathbb{Z}/n\mathbb{Z}} b_m x^{2^m},$$

$$L_2(x) = \sum_{p \in \mathbb{Z}/n\mathbb{Z}} c_p x^{2^p}.$$

Then we get

$$\sum_{m \in \mathbb{Z}/n\mathbb{Z}} b_m x^{2^m(2^s+1)} + \sum_{m \in \mathbb{Z}/n\mathbb{Z}} b_m a^{2^m} x^{2^{m+j}(2^q+1)}$$

$$= \sum_{l,p \in \mathbb{Z}/n\mathbb{Z}} c_p c_l^{2^r} x^{2^{l+r}+2^p} + L'(x). \tag{5.5}$$

On the left hand side of the identity (5.5) we have only items of the type $x^{2^m(2^s+1)}$, $x^{2^{m+j}(2^q+1)}$, with some coefficients. Therefore this must be true also for the right hand side of the identity.

We shall show that under some conditions on s, j, q, the equality above is satisfied only if $b_m = 0$ for every $m \in \mathbb{Z}/n\mathbb{Z}$. A contradiction.

If $b_m \neq 0$ for some m, then the coefficients of the items $x^{2^m(2^s+1)}$ and $x^{2^{m+j}(2^q+1)}$ are not zero on the left hand side of the identity (5.5) since $q \neq \pm s$. Hence this is also true for the right hand side of (5.5), that is,

$$c_m c_{m+s-r}^{2^r} \neq c_{m+s} c_{m-r}^{2^r}, \tag{5.6}$$

$$c_{m+j} c_{m+j+q-r}^{2^r} \neq c_{m+j+q} c_{m+j-r}^{2^r}. \tag{5.7}$$

The items of the type $x^{2^m+2^{m+j}}$ are missing in the left hand side of (5.5) when $j \neq \pm s, \pm q$. And we have no item of the kind $x^{2^{m+j}+2^{m+s}}$ in the left hand side of (5.5) when $j - s \neq \pm s, \pm q$, that is, $j \neq 2s, s \pm q$.

Thus, if these conditions are satisfied, then from the right hand side of (5.5) we get the following equalities with $c_m, c_{m+s-r}^{2^r}, c_{m+s}, c_{m-r}^{2^r}, c_{m+j}, c_{m+j-r}^{2^r}$:

$$c_m c_{m+j-r}^{2^r} = c_{m+j} c_{m-r}^{2^r}, \tag{5.8}$$

$$c_{m+j} c_{m+s-r}^{2^r} = c_{m+s} c_{m+j-r}^{2^r}. \tag{5.9}$$

Assume $c_{m+j-r}, c_{m+s-r} \neq 0$. If $c_{m-r} \neq 0$ then we get from (5.6), (5.8), (5.9):

$$c_m c_{m-r}^{-2^r} \neq c_{m+s} c_{m+s-r}^{-2^r},$$

$$c_m c_{m-r}^{-2^r} = c_{m+j} c_{m+j-r}^{-2^r},$$

$$c_{m+j} c_{m+j-r}^{-2^r} = c_{m+s} c_{m+s-r}^{-2^r},$$

and we come to an obvious contradiction. If $c_{m-r} = 0$ then from (5.8) and since $c_{m+j-r} \neq 0$ we get $c_m = 0$. But $c_{m-r} = c_m = 0$ contradicts (5.6). Therefore, either c_{m+j-r} or c_{m+s-r} equals 0.

Assume first that $c_{m+j-r} = 0$. Then from (5.7) we get $c_{m+j} \neq 0$; then from (5.8), (5.9) we get $c_{m+s-r} = c_{m-r} = 0$, that is in contradiction with (5.6). Therefore, $c_{m+j-r} \neq 0$.

Assume now that $c_{m+s-r} = 0$. Then from (5.6) we get $c_{m+s} \neq 0$; then from (5.9) we get $c_{m+j-r} = 0$. Then from (5.7) we get $c_{m+j} \neq 0$, which in its turn gives

$c_{m-r} = 0$ because of (5.8). Thus, we get $c_{m+s-r} = c_{m-r} = 0$ which contradicts (5.6).

Therefore, if $j \neq \pm s, \pm q, 2s, s \pm q$ then F is EA-inequivalent to quadratic power functions.

Using similar arguments we get below other conditions on s, q, j which are also sufficient.

Let $q \neq \pm s$ and $j \neq \pm s, \pm q, \pm s - q, -2q$. Then we have inequalities (5.6), (5.7), and the equality (5.8). Besides, we have no items of the kind $x^{2^{m+j+q}+2^m}$ in the left hand side of (5.5) when $j + q \neq \pm s, \pm q$, that is, $j \neq \pm s - q, -2q$. Thus, from (5.5) we get the following equality

$$c_m c_{m+j+q-r}^{2^r} = c_{m+j+q} c_{m-r}^{2^r}. \tag{5.10}$$

Let $c_{m+j+q-r}, c_{m+j-r} \neq 0$. If also $c_{m-r} \neq 0$ then we get from (5.7), (5.8), (5.10)

$$c_{m+j} c_{m+j-r}^{-2^r} \neq c_{m+j+q} c_{m+j+q-r}^{-2^r},$$

$$c_m c_{m-r}^{-2^r} = c_{m+j} c_{m+j-r}^{-2^r},$$

$$c_m c_{m-r}^{-2^r} = c_{m+j+q} c_{m+j+q-r}^{-2^r},$$

and we come to a contradiction. If $c_{m-r} = 0$ then it follows from (5.8) that $c_m = 0$. But $c_m = c_{m-r} = 0$ contradicts (5.6). Therefore, either $c_{m+j+q-r} = 0$ or $c_{m+j-r} = 0$.

If $c_{m+j-r} = 0$ then $c_{m+j}, c_{m+j+q-r} \neq 0$ by (5.7). Since $c_{m+j-r} = 0$ and $c_{m+j} \neq 0$ then it follows from (5.8) that $c_{m-r} = 0$. Since $c_{m+j+q-r} \neq 0$ and $c_{m-r} = 0$ then $c_m = 0$ by (5.10). But $c_{m-r} = c_m = 0$ contradicts (5.6).

If $c_{m+j+q-r} = 0$ then from (5.7) we get $c_{m+j+q}, c_{m+j-r} \neq 0$. Since $c_{m+j+q-r} = 0$ and $c_{m+j+q} \neq 0$ then $c_{m-r} = 0$ from (5.10). We have $c_m = 0$ from (5.8) since $c_{m+j-r} \neq 0$ and $c_{m-r} = 0$. But $c_m = c_{m-r} = 0$ contradicts (5.6).

Thus, if $j \neq \pm s, \pm q, \pm s - q, -2q$ then the function F is EA-inequivalent to power functions.

The proofs of the third and the fourth claim of the theorem are similar to those for the first and the second cases. That is why further we give only the sketch of the proofs. No items of the kind $x^{2^{m+j+q}+2^m}$ occur in the left hand side of (5.5) when $j \neq 2s - q, s, -q, s - 2q$. This way we get the following equality

$$c_{m+s} c_{m+j+q-r}^{2^r} = c_{m+j+q} c_{m+s-r}^{2^r}. \tag{5.11}$$

The equalities (5.9) and (5.11) are in contradiction with inequalities (5.6) and (5.7). Thus the condition $j \neq 2s - q, s, -q, s - 2q, s \pm q, 2s$ is sufficient for F to be EA-inequivalent to power functions. The same is true when we consider the equalities (5.10) and (5.11) with the condition $j \neq 2s - q, s, -q, s - 2q, \pm s - q, -2q$. $\square$proof

Corollary 17 [14] *The function of Corollary 16 is EA-inequivalent to power functions when $k \geq 4$.*

Proof The function F corresponds to the first case in the hypotheses of Theorem 25. Indeed, if $i = 1$ then in terms of Theorem 25 we have $j = k, q = k + s$

and the conditions $q \neq \pm s$, $j \neq \pm s, \pm q, s \pm q, 2s$ are equivalent to $k + s \neq \pm s$, $k \neq s, 3k - s, k + s, 2k - s, k + 2s, 2k, 2s$ which are satisfied since $k \geq 4$ and $\gcd(k, 3) = \gcd(s, 3k) = 1$. Hence, the function F is EA-inequivalent to power functions by Theorem 25. $\qquad\square$

Corollary 18 [14] *The functions of Theorem 24 are EA-inequivalent to power functions when $k \geq 3$.*

Proof The function F satisfies the conditions of Theorem 25. If $i = 1$ then $j = k$ and $q = 2k + s$. The conditions $q \neq \pm s$, $j \neq \pm s, \pm q, 2s, s \pm q$ are satisfied when $k \geq 3$ because k, s are odd, $n = 4k$, $\gcd(s, 4k) = 1$. $\qquad\square$

To prove CCZ-inequivalence of the APN binomials with Gold and Kasami functions we need the following observation. By definition, functions F and G from $\mathbb{F}_{2^n}$ to itself are CCZ-equivalent if and only if there exists an affine automorphism $\mathcal{L} = (L_1, L_2)$ of $\mathbb{F}_{2^n} \times \mathbb{F}_{2^n}$ such that

$$y = F(x) \Leftrightarrow L_2(x, y) = G(L_1(x, y)).$$

The function $L_1(x, F(x))$ has to be a permutation too. Indeed, suppose that there exists $x \neq x'$ such that $L_1(x, F(x)) = L_1(x', F(x'))$, then since $\mathcal{L}$ is a permutation, we would have $L_2(x, F(x)) \neq L_2(x', F(x'))$, a contradiction since $L_2(x, F(x)) = G(L_1(x, F(x)))$ and $L_2(x', F(x')) = G(L_1(x', F(x')))$. Note also that, conversely, if F and $\mathcal{L} = (L_1, L_2)$ are respectively a function and an affine automorphism such that the function $L_1(x, F(x))$ is a permutation, then the relation $L_2(x, F(x)) = G(L_1(x, F(x)))$ defines a function G which is CCZ-equivalent to F.

Theorem 26 [14] *Let n be a positive integer, let r, s, q be three nonzero elements of $\mathbb{Z}/n\mathbb{Z}$ and j an element of $\mathbb{Z}/n\mathbb{Z}$. Let a be a nonzero element of $\mathbb{F}_{2^n}$. Assume that $s \neq \pm q$ and one of the following two conditions is satisfied*

1) $j \notin \{-q, \pm s, \pm s - q, s - r, -r, s - r - q, -r - q\}$;
2) $j \notin \{-q, \pm s, \pm s - q, s + r, r, s + r - q, r - q\}$.

If $F(x) = x^{2^s + 1} + ax^{2^j(2^q + 1)}$ is CCZ-equivalent to the function $G(x) = x^{2^r + 1}$ with $\gcd(r, n) = 1$ then F and G are EA-equivalent.

Proof Suppose that $F(x)$ and $G(x)$ are CCZ-equivalent, that is, there exists an affine automorphism $\mathcal{L} = (L_1, L_2)$ of $\mathbb{F}_{2^n} \times \mathbb{F}_{2^n}$ such that $y = F(x) \Leftrightarrow L_2(x, y) = G(L_1(x, y))$. This implies then $L_1(x, F(x))$ is a permutation and $L_2(x, F(x)) = G(L_1(x, F(x)))$. Writing $L_1(x, y) = L(x) + L'(y)$ and $L_2(x, y) = L''(x) + L'''(y)$ gives

$$L''(x) + L'''(F(x)) = G[L(x) + L'(F(x))]. \qquad (5.12)$$

We can write

$$L(x) = b + \sum_{m \in \mathbb{Z}/n\mathbb{Z}} b_m x^{2^m}, \qquad (5.13)$$

$$L'(x) = b' + \sum_{m \in \mathbb{Z}/n\mathbb{Z}} b'_m x^{2^m}, \qquad (5.14)$$

$$L''(x) = b'' + \sum_{m \in \mathbb{Z}/n\mathbb{Z}} b''_m x^{2^m}, \tag{5.15}$$

$$L'''(x) = b''' + \sum_{m \in \mathbb{Z}/n\mathbb{Z}} b'''_m x^{2^m}, \tag{5.16}$$

$$b + b' = c. \tag{5.17}$$

$$b'' + b''' = c'. \tag{5.18}$$

We have

$$G[L(x) + L'(F(x))] = \left(L(x) + L'(x^{2^s+1} + ax^{2^j(2^q+1)}) \right)$$

$$\times \left(L(x) + L'(x^{2^s+1} + ax^{2^j(2^q+1)}) \right)^{2^r}$$

$$= (c + \sum_{m \in \mathbb{Z}/n\mathbb{Z}} b_m x^{2^m} + \sum_{m \in \mathbb{Z}/n\mathbb{Z}} b'_m x^{2^m(2^s+1)}$$

$$+ \sum_{m \in \mathbb{Z}/n\mathbb{Z}} a^m b'_m x^{2^{j+m}(2^q+1)})$$

$$\times (c^{2^r} + \sum_{m \in \mathbb{Z}/n\mathbb{Z}} b_m^{2^r} x^{2^{m+r}} + \sum_{m \in \mathbb{Z}/n\mathbb{Z}} b'^{2^r}_m x^{2^{m+r}(2^s+1)}$$

$$+ \sum_{m \in \mathbb{Z}/n\mathbb{Z}} a^{2^{r+m}} b'^{2^r}_m x^{2^{r+j+m}(2^q+1)})$$

$$= Q(x) + [\sum_{m,k \in \mathbb{Z}/n\mathbb{Z}} b_k b'^{2^r}_m x^{2^{m+r}(2^s+1)+2^k}$$

$$+ \sum_{m,k \in \mathbb{Z}/n\mathbb{Z}} a^{2^{r+m}} b_k b'^{2^r}_m x^{2^{r+j+m}(2^q+1)+2^k}$$

$$+ \sum_{m,k \in \mathbb{Z}/n\mathbb{Z}} b'_k b_m^{2^r} x^{2^{m+r}+2^k(2^s+1)}$$

$$+ \sum_{m,k \in \mathbb{Z}/n\mathbb{Z}} a^{2^k} b'_k b_m^{2^r} x^{2^{m+r}+2^{j+k}(2^q+1)}]$$

$$+ [\sum_{m,k \in \mathbb{Z}/n\mathbb{Z}} b'_k b'^{2^r}_m x^{2^{m+r}(2^s+1)+2^k(2^s+1)}$$

$$+ \sum_{m,k \in \mathbb{Z}/n\mathbb{Z}} a^{2^{r+m}} b'_k b'^{2^r}_m x^{2^{r+j+m}(2^q+1)+2^k(2^s+1)}$$

$$+ \sum_{m,k \in \mathbb{Z}/n\mathbb{Z}} a^{2^k} b'_k b'^{2^r}_m x^{2^{m+r}(2^s+1)+2^{j+k}(2^q+1)}$$

$$+ \sum_{m,k \in \mathbb{Z}/n\mathbb{Z}} a^{2^{r+m}+2^k} b'_k b'^{2^r}_m x^{2^{r+j+m}(2^q+1)+2^{j+k}(2^q+1)}],$$

where $Q(x)$ is a quadratic polynomial and in the second bracket we collect all items which potentially can have algebraic degree 4 (that is, whose exponents can have 2-weight equal to 4).

Obviously, all terms in the expression above whose exponents have 2-weight strictly greater than 2 must cancel because of identity (5.12).

If L' is a constant then F and G are obviously EA-equivalent and it proves the statement of the theorem. If the function L' is not a constant then there exists $m \in \mathbb{Z}/n\mathbb{Z}$ such that $b'_m \neq 0$. If $j \neq s-r$, $j \neq -r$, $j+q \neq s-r$ and $j+q \neq -r$ then $2^{r+j+m}(2^q+1) + 2^m(2^s+1)$ has 2-weight 4 and the items with this exponent have to vanish. We get

$$a^{2^{m+r}} b'^{2^r+1}_m + a^{2^{m+r}} b'_{m+r} b'^{2^r}_{m-r} = 0$$

and since $a \neq 0$, $b'_m \neq 0$ then $b'_{m+r}, b'_{m-r} \neq 0$ and

$$b'_m b'^{-2^r}_{m-r} = b'_{m+r} b'^{-2^r}_m. \tag{5.19}$$

If $j \neq s+r$, $j \neq r$, $j+q \neq s+r$ and $j+q \neq r$ then $2^{m+j}(2^q+1)+2^{m+r}(2^s+1)$ has 2-weight 4 and we again get (5.19).

Since $\gcd(r,n) = 1$ then applying this observation for $m+r$, $m+2r$,..., instead of m we get $b'_t \neq 0$ and

$$b'_m b'^{-2^r}_{m-r} = b'_{t+r} b'^{-2^r}_t \tag{5.20}$$

for all $t \in \mathbb{Z}/n\mathbb{Z}$.

Let us consider the sum

$$\sum_{m,k \in \mathbb{Z}/n\mathbb{Z}} b'_k b'^{2^r}_m x^{2^{m+r}(2^s+1)+2^k(2^s+1)}$$

from the last bracket. For any $k, m \in \mathbb{Z}/n\mathbb{Z}$, $k \neq m+r$, the items $b'_k b'^{2^r}_m x^{2^{m+r}(2^s+1)+2^k(2^s+1)}$ and $b'_{m+r} b'^{2^r}_{k-r} x^{2^k(2^s+1)+2^{m+r}(2^s+1)}$ differ and cancel pairwise because of (5.20). In the case $k = m+r$ the sum gives items with the exponents of 2-weight not greater than 2.

Considering the sum

$$\sum_{m,k \in \mathbb{Z}/n\mathbb{Z}} a^{2^{r+m}+2^k} b'_k b'^{2^r}_m x^{2^{r+j+m}(2^q+1)+2^{j+k}(2^q+1)}$$

we get that for any $k, m \in \mathbb{Z}/n\mathbb{Z}$, $k \neq m+r$, the items

$$a^{2^{r+m}+2^k} b'_k b'^{2^r}_m x^{2^{r+j+m}(2^q+1)+2^{j+k}(2^q+1)}$$

and

$$a^{2^{r+m}+2^k} b'_{r+m} b'^{2^r}_{k-r} x^{2^{j+k}(2^q+1)+2^{r+j+m}(2^q+1)}$$

differ and cancel pairwise because of (5.20) and in the case $k = m + r$ the sum gives items with the exponents of 2-weight not greater than 2.

Now we consider the sums

$$\sum_{m,k \in \mathbb{Z}/n\mathbb{Z}} a^{2^{r+m}} b'_k b'^{2^r}_m x^{2^{r+j+m}(2^q+1)+2^k(2^s+1)}$$

and

$$\sum_{m,k \in \mathbb{Z}/n\mathbb{Z}} a^{2^k} b'_k b'^{2^r}_m x^{2^{m+r}(2^s+1)+2^{j+k}(2^q+1)}.$$

For any $k, m \in \mathbb{Z}/n\mathbb{Z}$ the item $a^{2^{r+m}} b'_k b'^{2^r}_m x^{2^{r+j+m}(2^q+1)+2^k(2^s+1)}$ from the first sum cancels with the item $a^{2^{r+m}} b'_{m+r} b'^{2^r}_{k-r} x^{2^k(2^s+1)+2^{r+j+m}(2^q+1)}$ from the second sum and vice versa.

Thus the expression in the last bracket is quadratic and

$$G[L(x) + L'(F(x))]$$

$$= Q'(x) + \Big[\sum_{m,k \in \mathbb{Z}/n\mathbb{Z}} b_k b'^{2^r}_m x^{2^{m+r}(2^s+1)+2^k} \tag{5.21}$$

$$+ \sum_{m,k \in \mathbb{Z}/n\mathbb{Z}} a^{2^{r+m}} b_k b'^{2^r}_m x^{2^{r+j+m}(2^q+1)+2^k} \tag{5.22}$$

$$+ \sum_{m,k \in \mathbb{Z}/n\mathbb{Z}} b'_k b^{2^r}_m x^{2^{m+r}+2^k(2^s+1)} \tag{5.23}$$

$$+ \sum_{m,k \in \mathbb{Z}/n\mathbb{Z}} a^{2^k} b'_k b^{2^r}_m x^{2^{m+r}+2^{j+k}(2^q+1)} \Big], \tag{5.24}$$

where $Q'(x)$ is a quadratic function.

Because of (5.20) we can deduce, by denoting $b'_r b'^{-2^r}_0 = \lambda$, that $b'_{t+r} = \lambda b'^{2^r}_t$ for all t. Then, introducing μ such that $\lambda = \mu^{2^r-1}$, we deduce that $\mu b'_{t+r} = (\mu b'_t)^{2^r}$ for all t and then that $\mu b'_{t+1} = (\mu b'_t)^2$ (using that $\gcd(r, n) = 1$) and then $\mu b'_t = (\mu b'_0)^{2^t}$. This means that $\mu L'(x) = \mu b' + \text{tr}(\mu b'_0 x)$. Then obviously L' is not a permutation and since $L_1(x, F(x))$ is a permutation then L is not a constant. Thus $b_t \neq 0$ for some $t \in \mathbb{Z}/n\mathbb{Z}$.

Now consider the exponent $2^{j+m}(2^q + 1) + 2^m$. This exponent has 2-weight 3 since $j \neq -q$. Furthermore, since $s \neq \pm q$ and $j \notin \{\pm s, \pm s - q\}$ then this exponent differs from exponents in the first and the third sums (i.e. in (5.21) and (5.23)). Thus, considering the corresponding coefficients from the parts (5.22) and (5.24) we get the equality

$$b_m b'^{-2^r}_{m-r} = b'_m b'^{-2^r}_{m-r}$$

and because of (5.20) we get for all $t \in \mathbb{Z}/n\mathbb{Z}$

$$\lambda = b'_m b'^{-2^r}_{m-r} = b_t b'^{-2^r}_{t-r}.$$

Therefore, $\mu L(x) = \mu b + \text{tr}(\mu b_0 x)$ and $\mu[L(x) + L'(F(x))] = \mu b' + \mu b + \text{tr}(\mu b_0 x + \mu b'_0 F(x))$. Obviously the function $L(x) + L'(F(x))$ is not a permutation and that is a contradiction. Therefore, L' is constant and F and G are EA-equivalent. □

Corollary 19 [14] *The functions of Corollary 16 are CCZ-inequivalent to the Gold mappings when $k \geq 4$.*

Proof Assume that the Gold function x^{2^r+1}, $\gcd(r, n) = 1$, is CCZ-equivalent to F. Then by Corollary 17 and by Theorem 26 one of the conditions $s \neq \pm q$, $j \neq -q$, $j \neq \pm s$, $j \neq \pm s - q$, $j \neq s - r$, $j \neq -r$, $j + q \neq s - r$, $j + q \neq -r$, is not satisfied.

Let $i = 1$. Then in terms of Theorem 26 we have $q = k + s$, $j = k$. If $s = \pm q$ or $j = -q$ or $j = \pm s$ or $j = \pm s - q$ then we get a contradiction with $n = 3k \neq 0$ or $\gcd(s, k) = 1$. If $r = -j$ or $r = s - (j + q)$ then $\gcd(r, k) \neq 1$, again a contradiction. If $r = s - j$ or $r = -(j + q)$ then r is divisible by 3. Indeed, since $sk = 1 \bmod 3$ then $s \bmod 3 = k \bmod 3$ and $\pm(s - k) = 0 \bmod 3$. On the other hand, $r = s - j = s - k$ or $r = -(j + q) = n - (2k + s) = 3k - (2k + s) = k - s$. But $\gcd(r, 3k) = 1$, a contradiction. □

Corollary 20 [14] *The function F of Theorem 24 is CCZ-inequivalent to the Gold mappings when $k \geq 3$.*

Proof The proof for CCZ-inequivalence to the Gold functions is based on Corollary 18 and Theorem 26. Let $i = 1$, then $j = k$ and $q = 2k + s$ satisfy the conditions $q \neq \pm s$, $j \neq -q$, $j \neq \pm s$, $j \neq \pm s - q$, $j \neq s - r$, $j \neq -r$, $j + q \neq s - r$, $j + q \neq -r$ for any r satisfying $1 \leq r < n/2$ and $\gcd(r, n) = 1$. Indeed, the cases $q = \pm s$ or $j = -q$ or $j = \pm s$ or $j = \pm s - q$ are in contradiction with $\gcd(s, 4k) = 1$, $n = 4k$. If $k = s - r$ then it contradicts the fact that k is odd and $s - r$ is even. If $k = -r$ then it would contradict $\gcd(r, 4k) = 1$. If $3k + s = s - r$ then $3k = -r$ and $\gcd(r, k) \neq 1$, a contradiction. If $3k + s = -r$ then $s + r = k$ while s, r, k are odd. By Theorem 26 and Corollary 18 the function F is CCZ-inequivalent to x^{2^r+1}. □

Remark 15 Due to the recent work [32], where it is proven that two quadratic APN functions are CCZ-equivalent if and only if they are EA-equivalent, we can also get the results of Corollaries 19 and 20 directly from Corollaries 17 and 18.

Now we consider the case of Kasami functions.

Theorem 27 [14] *Let n be a positive integer grater than 4, let r, s, q, j be nonzero elements of $\mathbb{Z}/n\mathbb{Z}$ such that $\gcd(r, n) = 1$, $2 \leq r < n/2$, $s \notin \{\pm q, \pm 3q\}$, $q \neq \pm 3s$, $j \notin \{\pm s, \pm q, \pm s - q, s \pm q, -2q, s - 2q, 2s, 2s - q\}$. Then for $a \in \mathbb{F}_{2^n}^*$ the functions $F(x) = x^{2^s+1} + ax^{2^j(2^q+1)}$ and $K(x) = x^{4^r - 2^r + 1}$ are CCZ-inequivalent.*

Proof Let $G(x) = x^{2^r+1}$, $G'(x) = x^{2^{3r}+1}$. Suppose that $F(x)$ and $K(x)$ are CCZ-equivalent. Then, there exists an affine automorphism $\mathcal{L} = (L_1, L_2)$ of $\mathbb{F}_{2^n} \times \mathbb{F}_{2^n}$ such that $L_2(x, F(x)) = K(L_1(x, F(x)))$, which implies, by composition by G

$$G(L_2(x, F(x))) = G'(L_1(x, F(x))),$$

that is, writing again $L_1(x, y) = L(x) + L'(y)$ and $L_2(x, y) = L''(x) + L'''(y)$ and using the notations (5.14–5.18):

$$0 = G'[L(x) + L'(F(x))] + G[L''(x) + L'''(F(x))]$$

$$= \left(L(x) + L'(x^{2^s+1} + ax^{2^j(2^q+1)})\right)$$

$$\times \left(L(x) + L'(x^{2^s+1} + ax^{2^j(2^q+1)})\right)^{2^{3r}}$$

$$+ \left(L''(x) + L'''(x^{2^s+1} + ax^{2^j(2^q+1)})\right)$$

$$\times \left(L''(x) + L'''(x^{2^s+1} + ax^{2^j(2^q+1)})\right)^{2^r}$$

$$= [c + \sum_{m\in\mathbb{Z}/n\mathbb{Z}} b_m x^{2^m} + \sum_{m\in\mathbb{Z}/n\mathbb{Z}} b'_m x^{2^m(2^s+1)}$$

$$+ \sum_{m\in\mathbb{Z}/n\mathbb{Z}} a^m b'_m x^{2^{j+m}(2^q+1)}] \times [c^{2^{3r}} + \sum_{m\in\mathbb{Z}/n\mathbb{Z}} b_m^{2^{3r}} x^{2^{m+3r}}$$

$$+ \sum_{m\in\mathbb{Z}/n\mathbb{Z}} b'^{2^{3r}}_m x^{2^{m+3r}(2^s+1)} + \sum_{m\in\mathbb{Z}/n\mathbb{Z}} a^{2^{3r+m}} b'^{2^{3r}}_m x^{2^{3r+j+m}(2^q+1)}]$$

$$+ [c' + \sum_{m\in\mathbb{Z}/n\mathbb{Z}} b''_m x^{2^m} + \sum_{m\in\mathbb{Z}/n\mathbb{Z}} b'''_m x^{2^m(2^s+1)}$$

$$+ \sum_{m\in\mathbb{Z}/n\mathbb{Z}} a^m b'''_m x^{2^{j+m}(2^q+1)}] \times [c'^{2^r} + \sum_{m\in\mathbb{Z}/n\mathbb{Z}} b''^{2^r}_m x^{2^{m+r}}$$

$$+ \sum_{m\in\mathbb{Z}/n\mathbb{Z}} b'''^{2^r}_m x^{2^{m+r}(2^s+1)} + \sum_{m\in\mathbb{Z}/n\mathbb{Z}} a^{2^r+m} b'''^{2^r}_m x^{2^{r+j+m}(2^q+1)}]$$

$$= Q(x) + [\sum_{m,k\in\mathbb{Z}/n\mathbb{Z}} b_k b'^{2^{3r}}_m x^{2^{m+3r}(2^s+1)+2^k}$$

$$+ \sum_{m,k\in\mathbb{Z}/n\mathbb{Z}} a^{2^{3r+m}} b_k b'^{2^{3r}}_m x^{2^{3r+j+m}(2^q+1)+2^k}$$

$$+ \sum_{m,k\in\mathbb{Z}/n\mathbb{Z}} b'_k b^{2^{3r}}_m x^{2^{m+3r}+2^k(2^s+1)}$$

$$+ \sum_{m,k\in\mathbb{Z}/n\mathbb{Z}} a^{2^k} b'_k b^{2^{3r}}_m x^{2^{m+3r}+2^{j+k}(2^q+1)}$$

$$+ \sum_{m,k\in\mathbb{Z}/n\mathbb{Z}} b''_k b'''^{2^r}_m x^{2^{m+r}(2^s+1)+2^k}$$

$$+ \sum_{m,k\in\mathbb{Z}/n\mathbb{Z}} a^{2^r+m} b''_k b'''^{2^r}_m x^{2^{r+j+m}(2^q+1)+2^k}$$

$$+ \sum_{m,k\in\mathbb{Z}/n\mathbb{Z}} b_k''' b_m''^{2^r} x^{2^{m+r}+2^k(2^s+1)}$$

$$+ \sum_{m,k\in\mathbb{Z}/n\mathbb{Z}} a^{2^k} b_k''' b_m''^{2^r} x^{2^{m+r}+2^{j+k}(2^q+1)}]$$

$$+ [\sum_{m,k\in\mathbb{Z}/n\mathbb{Z}} b_k' b_m'^{2^{3r}} x^{2^{m+3r}(2^s+1)+2^k(2^s+1)}$$

$$+ \sum_{m,k\in\mathbb{Z}/n\mathbb{Z}} a^{2^{3r+m}} b_k' b_m'^{2^{3r}} x^{2^{3r+j+m}(2^q+1)+2^k(2^s+1)}$$

$$+ \sum_{m,k\in\mathbb{Z}/n\mathbb{Z}} a^{2^k} b_k' b_m'^{2^{3r}} x^{2^{m+3r}(2^s+1)+2^{j+k}(2^q+1)}$$

$$+ \sum_{m,k\in\mathbb{Z}/n\mathbb{Z}} a^{2^{3r+m}+2^k} b_k' b_m'^{2^{3r}} x^{2^{3r+j+m}(2^q+1)+2^{j+k}(2^q+1)}$$

$$+ \sum_{m,k\in\mathbb{Z}/n\mathbb{Z}} b_k''' b_m'''^{2^r} x^{2^{m+r}(2^s+1)+2^k(2^s+1)}$$

$$+ \sum_{m,k\in\mathbb{Z}/n\mathbb{Z}} a^{2^{r+m}} b_k''' b_m'''^{2^r} x^{2^{r+j+m}(2^q+1)+2^k(2^s+1)}$$

$$+ \sum_{m,k\in\mathbb{Z}/n\mathbb{Z}} a^{2^k} b_k''' b_m'''^{2^r} x^{2^{m+r}(2^s+1)+2^{j+k}(2^q+1)}$$

$$+ \sum_{m,k\in\mathbb{Z}/n\mathbb{Z}} a^{2^{r+m}+2^k} b_k''' b_m'''^{2^r} x^{2^{r+j+m}(2^q+1)+2^{j+k}(2^q+1)}],$$

where Q is quadratic and the items which can give algebraic degree 4 are collected in the second brackets.

The exponents of the type $2^{3r+j+m+q} + 2^{3r+j+m} + 2^{k+s} + 2^k$ have 2-weight 4 if $k \notin \{3r+j+m, 3r+j+m+q, 3r+j+m+q-s, 3r+j+m-s\}$. Besides, these exponents cannot be equal to any exponent of the type $(2^s+1)(2^l+2^{l'})$ when $q \neq \pm s, \pm 3s$ and of the type $(2^q+1)(2^l+2^{l'})$ when $s \neq \pm q, \pm 3q$. Since all terms with exponents of 2-weight 4 should vanish we obtain

$$b_k' b_m'^{2^{3r}} + b_{m+3r}' b_{k-3r}'^{2^{3r}} = b_k''' b_{m+2r}'''^{2^r} + b_{m+3r}''' b_{k-r}''^{2^r} \tag{5.25}$$

for $m, k \in \mathbb{Z}/n\mathbb{Z}, k \notin \{3r+j+m, 3r+j+m+q, 3r+j+m+q-s, 3r+j+m-s\}$.

Equality (5.25) is also true for the cases $k \in \{3r+j+m, 3r+j+m+q, 3r+j+m+q-s, 3r+j+m-s\}$ when $s \neq \pm q, j \notin \{\pm s, \pm q, \pm s-q, s+q, -2q, s-2q, 2s, 2s-q\}$.

Indeed, let us consider the items with the exponents $2^{3r+j+m}(2^q+1)+2^{j+k}(2^q+1)$ for $k \in \{3r+j+m, 3r+j+m+q\}$.

If $k = 3r+j+m$ then

$$2^{3r+j+m}(2^q+1) + 2^{j+k}(2^q+1) = 2^{3r+j+m}(2^q+2^0+2^{j+q}+2^j)$$

and has 2-weight 4 since $j \neq \pm q$ and it differs from exponents of the type $(2^s + 1)(2^l + 2^{l'})$ since $j \notin \{\pm s - q, \pm s\}$ and of the type $2^l(2^q + 1) + 2^{l'}(2^s + 1)$ since $j \notin \{\pm s - q, \pm s, -2q, \pm q\}$.

If $k = 3r + j + m + q$ then

$$2^{3r+j+m}(2^q + 1) + 2^{j+k}(2^q + 1) = 2^{3r+j+m}(2^q + 2^0 + 2^{j+2q} + 2^{j+q})$$

has 2-weight 4 since $j \notin \{-q, -2q\}$ and it differs from exponents of the type $(2^s + 1)(2^l + 2^{l'})$ because $j \notin \{\pm s - q, \pm s\}$ and of the type $2^l(2^q + 1) + 2^{l'}(2^s + 1)$ since $j \notin \{\pm s - q, \pm s, \pm q, -2q\}$.

For $k \in \{3r + j + m + q - s, 3r + j + m - s\}$ we can consider $2^{m+3r}(2^s + 1) + 2^k(2^s + 1)$. Indeed, if $k = 3r + j + m + q - s$ then

$$2^{m+3r}(2^s + 1) + 2^k(2^s + 1) = 2^{m+3r}(2^s + 2^0 + 2^{j+q} + 2^{j+q-s})$$

and has 2-weight 4 since $j \notin \{s - q, 2s - q, -q\}$ and it differs from exponents of the type $(2^q + 1)(2^l + 2^{l'})$ since $j \notin \{-2q, s, -2q + s\}$ and $2^l(2^q + 1) + 2^{l'}(2^s + 1)$ since $j \notin \{-2q, s, -2q + s, \pm s - q, -q, 2s - q\}$.

If $k = 3r + j + m - s$ then

$$2^{m+3r}(2^s + 1) + 2^k(2^s + 1) = 2^{m+3r}(2^s + 2^0 + 2^j + 2^{j-s})$$

has 2-weight 4 since $j \notin \{s, 2s\}$ and it differs from exponents of the type $(2^q + 1)(2^l + 2^{l'})$ because $j \notin \{\pm q + s, \pm q\}$ and of the type $2^l(2^q + 1) + 2^{l'}(2^s + 1)$ since $j \notin \{\pm q + s, \pm q, \pm s, 2s\}$.

Without loss of generality we can assume that L, L', L'', L''' are linear (since changing the constant terms in these affine mappings results only in a change of the polynomial $Q(x)$ above) and let $L' \neq 0$. The equalities (5.25) imply

$$(L'''(x))^{2^r+1} + (L'(x))^{2^{3r}+1} = C(x) \tag{5.26}$$

for some linear function $C(x)$. Besides, it must hold that

$$\ker(L''') \cap \ker(L') = \{0\} \tag{5.27}$$

since otherwise the system of equations

$$L(x) + L'(y) = 0$$
$$L''(x) + L'''(y) = 0$$

has solutions different from $(0, 0)$ which is not possible since $\mathcal{L}$ is a permutation.

For any element u, derivating equality (5.26) we get

$$L'''(u)^{2^r} L'''(x) + L'''(u)L'''(x)^{2^r}$$

$$+ L'(u)^{2^{3r}} L'(x) + L'(u)L'(x)^{2^{3r}} = 0 \tag{5.28}$$

We want to show first that L' and L''' have to be bijective. Assume on the contrary that L' is not bijective. Then there exists an element $u_0 \neq 0$ such that $L'(u_0) = 0$, and due to equality (5.27) $L'''(u_0) \neq 0$. We get for all x that

$$L'''(u_0)^{2^r} L'''(x) + L'''(u_0)L'''(x)^{2^r} = 0.$$

And it follows that

$$L'''(x) = 0 \text{ or } L'''(x) = L'''(u_0),$$

where we used that $\gcd(2^r - 1, 2^n - 1) = 1$. Thus there exists an element d such that

$$L'''(x) = L'''(u_0)\text{tr}(dx).$$

If we plug this into equality (5.28) we get

$$L'(u)^{2^{3r}} L'(x) + L'(u)L'(x)^{2^{3r}} = 0$$

for all x and any a. This implies that $L'(x) = 0$ or

$$L'(x)^{2^{3r}-1} = L'(u)^{2^{3r}-1}$$

which, as $\gcd(3r, n) = 3$, means that

$$L'(x) = L'(u)\gamma$$

where $\gamma \in \mathbb{F}_{2^3}$. In particular we have

$$\dim(\text{im}(L')) \leq 3$$

and therefore we have $\dim(\ker(L')) \geq n - 3$. As $\dim(\ker(L''')) = n - 1$ for $n > 4$ the two kernel intersect, a contradiction. Thus, L' must be bijective.

Now assume that L''' is not bijective. Then there exists u_1 such that $L'''(u_1) = 0$ and $L'(u_1) \neq 0$. We get, again

$$L'(u_1)^{2^{3r}} L'(x) + L'(u_1)L'(x)^{2^{3r}} = 0$$

which, using the same arguments as above, contradicts the condition that L' is bijective. We conclude that L''' is bijective.

Now we denote $A = L''' \circ L'^{-1}$, which is again a bijective linear mapping. By replacing x by $L'^{-1}(x)$ and u by $L'^{-1}(u)$ in (5.28) we obtain

$$A(u)^{2^r} A(x) + A(u)A(x)^{2^r} + u^{2^{3r}}x + ux^{2^{3r}} = 0$$

and for $u \in \mathbb{F}_{2^3}$ we see that for all $x \in \mathbb{F}_{2^3}$ we get

$$A(u)^{2^r} A(x) + A(u)A(x)^{2^r} = 0$$

which is equivalent to $A(x) = 0$ or $A(x) = A(u)$ which is impossible since A is a bijection. This contradiction shows that the functions that the function L' is constant, and therefore the functions F and K are EA-equivalent. But this is also impossible because EA-equivalence preserves the algebraic degree. Indeed, F is quadratic while K is not when $2 \leq r < n/2$. Thus the functions F and K are CCZ-inequivalent. $\square$

Corollary 21 [14] *The functions of Corollary 16 are CCZ-inequivalent to the Kasami mappings when $k \geq 4$.*

Proof Let $i = 1$, then in terms of Theorem 27 we have $j = k, q = k + s$ for the function F of Corollary 16 and one can easily check that j, k, s here satisfy all conditions of Theorem 27 since $n = 3k, k \geq 4$ and $\gcd(s, 3k) = 1$. $\square$

Corollary 22 [14] *The function F of Theorem 24 is CCZ-inequivalent to the Kasami mappings when $k \geq 3$.*

Proof Let $i = 1$, then in terms of Theorem 27 we have $j = k, q = 2k + s$ for the function F of Theorem 24 and one can easily check that j, k, s here satisfy all conditions of Theorem 27 because k, s are odd, $n = 4k, k \geq 3$ and $\gcd(s, 4k) = 1$. $\square$

Proposition 21 [14] *When $k \geq 2$ the functions of Corollary 16 and Theorem 24 are CCZ-inequivalent to the inverse and Dobbertin mappings.*

Proof For any quadratic APN mapping F the number $2^{\lfloor \frac{n+1}{2} \rfloor}$ divides all the values in the Walsh spectrum of F (see [21, 31]). It is shown in [29] that the inverse function does not have this property. Besides, it is proven in [18] that $2^{\frac{2n}{5}+1}$ cannot be a divisor of all the values in the Walsh spectrum of a Dobbertin function. Since the extended Walsh spectrum of a function is invariant under CCZ-equivalence then F is CCZ-inequivalent to the inverse and Dobbertin functions. $\square$

Corollaries 19–22 and Proposition 21 let us state the following theorem.

Theorem 28 [14] *When $n \geq 12$ and n is even the functions of Corollary 16 and Theorem 24 are CCZ-inequivalent to all known power APN functions. When n is odd (and $n \geq 12$) these functions are CCZ-inequivalent to the Gold, Kasami, inverse and Dobbertin mappings.* $\square$

Theorem 28 implies that for $n = 12, 20, 24$ the introduced APN binomials are CCZ-inequivalent to all power functions. We conjecture that the functions from Corollary 16 and Theorem 24 are CCZ-inequivalent to any power function when $n \geq 12$.

5.3 Classes of APN Trinomials and Hexanomials

As shown in [23], one of the ways to construct APN polynomials is to consider quadratic hexanomials of the type

$$F(x) = x(Ax^2 + Bx^q + Cx^{2q}) + x^2(Dx^q + Ex^{2q}) + Gx^{3q} \tag{5.29}$$

over $\mathbb{F}_{2^{2m}}$ with $q = 2^m$. These polynomials are good candidates for being differentially 4-uniform, and potentially APN. This approach gave new examples of quadratic APN functions over $\mathbb{F}_{2^6}$ and $\mathbb{F}_{2^8}$ which are CCZ-inequivalent to power functions [23]. Besides, as we are going to see further, the infinite family of APN hexanomials (4) of Table 2.6 is based on construction (5.29) and it leads also to the family of APN trinomials (3).

Below we suggest natural generalizations of the method from [23], but first recall the arguments leading to the construction (5.29). Let a function F be defined by (5.29). Since F is quadratic then in order to determine its differential uniformity it is enough to know the numbers of solutions of the linear equations $F(x + a) + F(x) + F(a) = 0$ for all nonzero elements a of $\mathbb{F}_{2^{2m}}$. We get, for some expressions $a_1, \cdots, a_4, b_1, \cdots, b_3, c_1, \cdots, c_3$ depending on a:

$$f_1 = F(x + a) + F(x) + F(a) = a_1 x + a_2 x^2 + a_3 x^q + a_4 x^{2q} = 0,$$

$$f_2 = a_2^q f_1 + a_4 f_1^q = b_1 x + b_2 x^2 + b_3 x^q = 0,$$

$$f_3 = b_3^2 f_1 + a_3 b_3 f_2 + a_4 f_2^2 = c_1 x + c_2 x^2 + c_3 x^4 = 0.$$

Hence, if either c_1, c_2 or c_3 is different from 0 then F can have differential uniformity at most 4. In practice this condition on coefficients is very important. Indeed, the construction (5.29) gives, up to addition of affine functions, all quadratic functions on the field $\mathbb{F}_{2^4}$ while we have checked by running a computer that only about 3/4 of them have differential uniformity which is less or equal to 4. For the field F_{2^6} only about 18/41 of all functions generated by (5.29) have differential uniformity at most 4.

Let us now consider the construction

$$F'(x) = x(Ax^2 + Bx^4 + Cx^q + Dx^{2q} + Ex^{4q})$$
$$+ x^2(Gx^4 + Hx^q + Ix^{2q} + Jx^{4q})$$
$$+ x^4(Kx^q + Lx^{2q} + Mx^{4q}) + x^q(Nx^{2q} + Px^{4q}) + Qx^{2q+4q}.$$

For the function F' and for any nonzero elements a of $\mathbb{F}_{2^{2m}}$ we get

$$f_1' = F(x + a) + F(x) + F(a)$$
$$= a_1' x + a_2' x^2 + a_3' x^4 + a_4' x^q + a_5' x^{2q} + a_6' x^{4q} = 0,$$

$$f_2' = a_3'^q f_1' + a_6' f_1'^q$$
$$= b_1' x + b_2' x^2 + b_3' x^4 + b_4' x^q + b_5' x^{2q} = 0,$$

$$f_3' = b_3' f_1' + a_3' f_2'$$
$$= c_1' x + c_2' x^2 + c_3' x^q + c_4' x^{2q} + c_5' x^{4q} = 0,$$

$$f_4' = c_5'^q f_2' + b_3' f_3'^q = d_1' x + d_2' x^2 + d_3' x^q + d_4' x^{2q} = 0.$$

Thus, if some of the coefficients d_1', d_2', d_3', d_4' are different from 0 then we see that f_4' has the same form as f_1. Therefore, applying Dillon's method to $f_1 = f_4'$, we get that if some of the coefficients d_1', d_2', d_3', d_4', and some of the coefficients c_1, c_2, c_3 (associated to $f_1 = f_4'$) are different from 0 then F' is differentially 4-uniform. Obviously, the probability that we can prove this way that the function F' is differentially 4-uniform is less than in the case of construction (5.29) since we have an additional condition (on coefficients d_1', d_2', d_3', d_4'). And actually, we checked by a computer that only about $1/4$ of the quadratic functions on the field $\mathbb{F}_{2^6}$ are differentially 4-uniform (while all of them have the same form as F', up to addition of affine functions).

Clearly, construction (5.29) can be further generalized. For any i we denote

$$F^{(i)}(x) = \sum_{0 \le t < j \le i} a_{tj} x^{2^t + 2^j} + \sum_{0 \le t, j \le i} b_{tj} x^{2^t + 2^j q} + \sum_{0 \le t < j \le i} c_{tj} x^{q(2^t + 2^j)}$$

and consider $F^{(i)}$ over $\mathbb{F}_{2^{2m}}$ with $m \ge i + 1$. Obviously, the cases $i = 1, 2$ correspond to the functions F and F'. For arbitrary i, using induction, we get that, under a condition on the coefficients translating that no relation obtained in the process completely vanishes, the function $F^{(i)}$ is differentially 4-uniform. Note that all quadratic functions have the form $F^{(i)}(x)$ for $i = m - 1$. But clearly, with increasing i the probability that we can prove this way that $F^{(i)}$ is differentially 4-uniform decreases since the number of conditions on the coefficients grows. Nevertheless, we exhibit in the next section two subcases where these constructions succeed in providing differentially 4-uniform polynomials, and we can even deduce two new infinite classes of quadratic APN functions.

Note that functions whose nonzero derivatives are 2^k-to-1 mappings (i.e. reach any value either 0 or 2^k times) are studied in [12]. The simplest examples of such functions over $\mathbb{F}_{2^n}$ are $x^{2^i + 1}$ when $\gcd(i, n) = k$. The following theorems give new classes of such functions, which are differentially 4-uniform when $k = 2$ and APN when $k = 1$.

Theorem 29 [11] *Let m and i be any positive integers, $q = 2^m$, $n = 2m$, $\gcd(i, m) = k$ and $c, b \in \mathbb{F}_{2^n}$ be such that $c^{q+1} = 1$, $c \notin \{\lambda^{(2^i + 1)(q - 1)}, \lambda \in \mathbb{F}_{2^n}\}$, $cb^q + b \ne 0$. Then all the nonzero derivatives of the function*

$$F(x) = x^{2^{2i} + 2^i} + bx^{q+1} + cx^{q(2^{2i} + 2^i)}$$

are 2^k-to-1 mappings of $\mathbb{F}_{2^n}$.
Such vectors b, c do exist if and only if $\gcd(2^i + 1, q + 1) \ne 1$. For m odd, this is equivalent to saying that i is odd.

Proof Since F is quadratic, then for any nonzero a in $\mathbb{F}_{2^n}$ the function $F(x+a)+F(x)$ is 2^k-to-1 if and only if the equation

$$f_1 = F(x + a) + F(x) + F(a) = ba^q x + a^{2^{2i}} x^{2^i}$$
$$+ a^{2^i} x^{2^{2i}} + bax^q + ca^{2^{2i} q} x^{2^i q} + ca^{2^i q} x^{2^{2i} q} = 0$$

has 2^k solutions. This equation implies

$$f_1 + cf_1^q = (cb^q + b)(ax^q + a^q x) = 0$$

and, since $cb^q + b \neq 0$, then $ax^q + a^q x = 0$, i.e. $\left(\frac{x}{a}\right)^q = \frac{x}{a}$, and therefore $x = au$, $u \in \mathbb{F}_q$. The equation $f_1 = 0$ becomes

$$(a^{2^{2i}+2^i} + ca^{q(2^{2i}+2^i)})(u^{2^i} + u^{2^{2i}}) = 0.$$

The condition $c \notin \{\lambda^{(2^i+1)(q-1)}, \lambda \in \mathbb{F}_{2^n}\}$ implies that $a^{2^{2i}+2^i} + ca^{q(2^{2i}+2^i)} \neq 0$. Hence we have then $u^{2^i} + u^{2^{2i}} = 0$, that is, $u + u^{2^i} = 0$, i.e. $u = 0$ or $u^{2^i-1} = 1$. Since $\gcd(i,m) = k$ then $\gcd(2^i - 1, 2^m - 1) = 2^k - 1$ and $F(x+a) + F(x)$ is 2^k-to-one for any nonzero a.

Vectors c, b satisfying the hypotheses do exist if and only if there exists j such that $\alpha^{j(q-1)} \notin \{\lambda^{(2^i+1)(q-1)}, \lambda \in \mathbb{F}_{2^n}\}$, where α is a primitive element of F_{2^n} (we can take $b = 1$). Hence, the vectors c, b exist if and only if $(2^i+1) \cdot \mathbb{Z}/(q+1)\mathbb{Z} \neq \mathbb{Z}/(q+1)\mathbb{Z}$, that is, $\gcd(2^i + 1, q + 1) \neq 1$. For m odd, this is clearly equivalent to saying that i is odd. $\square$

Clearly, for k equal 1 and 2, Theorem 29 gives differentially 2- and 4-uniform functions respectively.

Corollary 23 [11] *Let m be any positive integer, $q = 2^m$, $n = 2m$, i be such that $\gcd(i,m) = 1$ and $\gcd(2^i + 1, q + 1) \neq 1$, and let $c, b \in \mathbb{F}_{2^n}$ be such that $c^{q+1} = 1$, $c \notin \{\lambda^{(2^i+1)(q-1)}, \lambda \in \mathbb{F}_{2^n}\}$, $cb^q + b \neq 0$. Then the function*

$$F(x) = x^{2^{2i}+2^i} + bx^{q+1} + cx^{q(2^{2i}+2^i)}$$

is APN on $\mathbb{F}_{2^n}$.

Theorem 30 [11] *Let m and i be any positive integers, $q = 2^m$, $n = 2m$, $\gcd(i,m) = k$, and $c, s \in \mathbb{F}_{2^n}$ be such that $s \notin \mathbb{F}_q$. If the equation*

$$x^{2^i+1} + cx^{2^i} + c^q x + 1 = 0$$

has no solution x such that $x^{q+1} = 1$, and in particular if the polynomial $X^{2^i+1} + cX^{2^i} + c^q X + 1$ is irreducible over $\mathbb{F}_{2^n}$, then all the nonzero derivatives of the function

$$F(x) = x(x^{2^i} + x^q + cx^{2^i q}) + x^{2^i}(c^q x^q + sx^{2^i q}) + x^{(2^i+1)q}$$

are 2^k-to-1 mappings of $\mathbb{F}_{2^n}$.

Proof As above, for any nonzero a, the function $F(x+a) + F(x)$ is 2^k-to-1 if and only if the equation $F(x+a) + F(x) + F(a) = 0$ has 2^k solutions.

We have

$$F(x+a) + F(x) + F(a) = (a^{2^i} + a^q + ca^{2^i q})x$$

$$+(a + c^q a^q + sa^{2^i q})x^{2^i} + (a + c^q a^{2^i} + a^{2^i q})x^q$$

$$+(ca + sa^{2^i} + a^q)x^{2^i q}$$

and

$$(F(x+a) + F(x) + F(a))^q = (a^{2^i q} + a + c^q a^{2^i})x^q$$

$$+(a^q + ca + s^q a^{2^i})x^{2^i q} + (a^q + ca^{2^i q} + a^{2^i})x$$
$$+(c^q a^q + s^q a^{2^i q} + a)x^{2^i}.$$

The sum of these two expressions equals $(s + s^q)(a^{2^i q} x^{2^i} + a^{2^i} x^{2^i q})$. Hence, since $s + s^q \neq 0$ then $F(x + a) + F(x) + F(a) = 0$ implies $ax^q + a^q x = 0$ and therefore $x = au$, $u \in \mathbb{F}_q$. Replacing x by au, we get

$$F(x + a) + F(x) + F(a)$$
$$= (u^{2^i} a^{2^i} + ua^q + cu^{2^i} a^{2^i q})a + (ua + c^q ua^q + su^{2^i} a^{2^i q})a^{2^i}$$
$$+(ua + c^q u^{2^i} a^{2^i} + u^{2^i} a^{2^i q})a^q + (cua + su^{2^i} a^{2^i} + ua^q)a^{2^i q}$$
$$= (u + u^{2^i})(a^{2^i+1} + a^{(2^i+1)q} + ca^{2^i q+1} + c^q a^{2^i+q}).$$

The equation $u + u^{2^i} = 0$ has 2^k solutions. We deduce that $F(x + a) + F(x)$ is 2^k-to-1 if the equation $x^{2^i+1} + x^{(2^i+1)q} + cx^{2^i q+1} + c^q x^{2^i+q} = 0$ admits no nonzero solution or, equivalently, the equation $x^{(2^i+1)(q-1)} + cx^{2^i(q-1)} + c^q x^{q-1} + 1 = 0$ has no solutions, or in other words, if the equation

$$y^{2^i+1} + cy^{2^i} + c^q y + 1 = 0$$

has no solution y such that $y^{q+1} = 1$. This happens (for instance) when the polynomial $X^{2^i+1} + cX^{2^i} + c^q X + 1$ is irreducible over $\mathbb{F}_{2^n}$.   □

Obviously, for the special case $k = 2$, Theorem 30 gives differentially 4-uniform functions and for $k = 1$, it gives a class of APN functions.

Corollary 24 [11] *Let m be any positive integer, $q = 2^m$, $n = 2m$, i be such that $\gcd(i, m) = 1$, and $c, s \in \mathbb{F}_{2^n}$ be such that $s \notin \mathbb{F}_q$. If the equation*

$$x^{2^i+1} + cx^{2^i} + c^q x + 1 = 0$$

has no solution x such that $x^{q+1} = 1$, and in particular if the polynomial $X^{2^i+1} + cX^{2^i} + c^q X + 1$ is irreducible over $\mathbb{F}_{2^n}$, then the function

$$F(x) = x(x^{2^i} + x^q + cx^{2^i q}) + x^{2^i}(c^q x^q + sx^{2^i q}) + x^{(2^i+1)q}$$

is APN on $\mathbb{F}_{2^n}$.

We checked with a computer that for $i = 1$, and at least for all even n, $6 \leq n \leq 1000$, not divisible by 3, there always exist elements $c \in \mathbb{F}_{2^n}$ for which the polynomial $X^{2^i+1} + cX^{2^i} + c^q X + 1$ is irreducible over $\mathbb{F}_{2^n}$. In case n is divisible by 6, $6 \leq n \leq 1000$, such elements exist at least for 140 out of 166 checked fields. We also checked that for $6 \leq n \leq 26$ the number of elements c for which the polynomial is irreducible is in average $3/10$-th of all elements. Moreover, as was proved recently in [3], elements c satisfying the conditions of Theorem 30 (and Corollary 24 in particular) always exist.

Inequivalence with Power Functions Dobbertin APN functions have Walsh spectra which are different from Walsh spectra of quadratic APN functions (see [17, 25, 31]). Since the extended Walsh spectrum of a function is invariant under CCZ-equivalence then we can make the following conclusion.

Proposition 22 [11] *For any positive n the functions of Corollaries 23 and 24 are CCZ-inequivalent to Dobbertin APN functions.*

We conjecture that the functions of Corollaries 23 and 24 are CCZ-inequivalent to all power functions for $n \geq 6$. In [23, 28] some invariants for CCZ-equivalence are presented in terms of group algebras and coding theory. We use one of them to prove the following proposition.

Proposition 23 [11] *At least some of the functions of Corollaries 23 and 24 are CCZ-inequivalent to power functions on $\mathbb{F}_{2^n}$ with $n = 6, 8, 10$.*

Let $G = \mathbb{F}_2[\mathbb{F}_{2^n} \times \mathbb{F}_{2^n}]$ be the group algebra of $\mathbb{F}_{2^n} \times \mathbb{F}_{2^n}$ over $\mathbb{F}_2$. It consists of the formal sums

$$\sum_{g \in G} a_g g$$

where $a_g \in \mathbb{F}_2$. If S is a subset of $\mathbb{F}_{2^n} \times \mathbb{F}_{2^n}$ then it can be identified with the element $\sum_{s \in S} s$ of G. The dimensions of the ideal of G generated by the graph G_F of F is called the Γ-rank of F. According to [23], the Γ-rank is CCZ-invariant.

For $n = 6$ we checked hundreds of functions of Corollaries 23 and 24, and all of them have Γ-ranks equal 1146 (take, for instance, $b = 1$ and $c = \alpha^{q-1}$ for the functions of Corollary 23, and $c = s = \alpha$ for the functions of Corollary 24 with α a primitive element of $\mathbb{F}_{2^6}^*$), while the only APN power function x^3 on $\mathbb{F}_{2^6}$ has the Γ-rank 1104. We further checked with a computer using coding approach from [9] that the functions of Corollaries 23 and 24 are CCZ-inequivalent to all APN power functions and all APN binomials up to $n = 10$.

Remark 16 When the APN trinomials and hexanomials of this section were first constructed, the only other known classes of APN functions were power APN functions and APN binomials (1–2) of Table 2.6. The APN multinomials (11) of Table 2.6 appeared in [5] a very shortly after the introduction of APN trinomials and hexanomials in [11] without giving any proof of difference with these families. We also cannot say much about the relation between the families of trinomials and hexanomials except that there are fields where the hexanomials (4) are defined while the trinomials (3) and the multinomials (11) are not. In any case, as proven in [20], all three families of APN functions (3), (4) and (11) are particular cases of a general construction of APN functions.

5.4   APN and AB Polynomials $x^3 + \text{tr}_n(x^9)$

In this section we present the family of APN and AB functions $x^3 + \text{tr}_n(x^9)$ over $\mathbb{F}_2^n$ constructed in [16]. This functions served as the first example of APN and AB polynomials CCZ-inequivalent to power functions whose all coefficients were in $\mathbb{F}_2$. Moreover it is still the only family of APN and AB polynomials CCZ-inequivalent to power functions which is defined for all n (recall that in case of power APN and AB functions only the Gold function x^3 posses this property).

We give a new approach for constructing quadratic APN functions and using it we construct a class of quadratic APN polynomials with coefficients in $\mathbb{F}_2$. We prove that the function $F(x) = x^3 + \text{tr}_n(x^9)$ is APN over $\mathbb{F}_{2^n}$ for any n, and that for all $n \geq 7$ it is CCZ-inequivalent to the Gold functions (and, hence, EA-inequivalent to all power functions since Gold functions are the only quadratic APN power functions), to the inverse and Dobbertin functions. Obviously, this function is AB for all odd n. We conjecture that for $n \geq 7$ the function F is CCZ-inequivalent to any power function. This conjecture is confirmed for the case $7 \leq n \leq 10$. Further we show that applying CCZ-equivalence to quadratic APN functions, it is possible to construct classes of nonquadratic APN mappings CCZ-inequivalent to power functions.

In the theorem below we give a general approach for constructing new quadratic APN functions from known ones.

Theorem 31   [16] *Let F be a quadratic APN function from $\mathbb{F}_{2^n}$ to itself, let f be a quadratic Boolean function on $\mathbb{F}_{2^n}$ and*

$$\varphi_F(x,a) = F(x) + F(x + a) + F(a) + F(0),$$
$$\varphi_f(x,a) = f(x) + f(x + a) + f(a) + f(0).$$

Then the function $F(x) + f(x)$ is APN if for every nonzero $a \in \mathbb{F}_{2^n}$ there exists a linear Boolean function ℓ_a satisfying the conditions

1) $\varphi_f(x,a) = \ell_a(\varphi_F(x,a))$,
2) *if $\varphi_F(x,a) = 1$ for some $x \in \mathbb{F}_{2^n}$ then $\ell_a(1) = 0$.*

Proof   Since the function $F(x) + f(x)$ is quadratic, it is APN if and only if, for every nonzero $a \in \mathbb{F}_{2^n}$, the equation $\varphi_F(x,a) + \varphi_f(x,a) = 0$ admits at most two solutions (see e.g. [19]). According to the hypothesis on ℓ_a, a solution to this equation must be such that $\varphi_f(x,a) = 0$ and therefore such that $\varphi_F(x,a) = 0$. Then, F being quadratic APN, this equation admits at most two solutions.   □

The same principle as in Theorem 31 allows generating a large variety of differentially 4-uniform functions from APN functions as it is shown in the proposition below.

Proposition 24   [16] *For any APN function F the following functions are differentially 4-uniform*

1) $F(x) + \text{tr}_n(G(x))$ for any function G;
2) $F \circ A$ and $A \circ F$ for any affine function A which is 2-to-1.

Remark 17 [16] *Note that, in the situation of Theorem 31, a linear function l_a satisfying $\varphi_f(x,a) = l_a(\varphi_F(x,a))$ always exists. This is due to the fact that, by the assumption F is APN and then the kernel of $\varphi_F(x,a)$ equals $\{0,a\}$. This set is always a subset of the kernel of $\varphi_f(x,a)$, which is indeed the necessary and sufficient condition for the existence of l_a.*   □

A direct consequence of Theorem 31 is that, if F is APN and if ℓ is a linear form such that $\ell(1) = 0$, then the function $F(x) + \ell(F(x))$ is APN. But this function is affine equivalent to F since it is equal to $L \circ F$ where $L(x) = x + \ell(x)$, and the condition that $\ell(1) = 0$ is equivalent to saying that L is a permutation.

We give now an example where Theorem 31 leads to a function which is CCZ-inequivalent to the original function F.

Corollary 25 [16] *Let n be any positive integer. Then the function $x^3 + \text{tr}_n(x^9)$ is APN on $\mathbb{F}_{2^n}$.*

Proof We can apply Theorem 31 with

$$F(x) = x^3,$$

$$\varphi_F(x,a) = a^2 x + ax^2,$$

$$f(x) = \text{tr}_n(x^9),$$

$$\varphi_f(x,a) = \text{tr}_n(a^8 x + ax^8),$$

$$\ell_a(y) = \text{tr}_n(a^6 y + a^3 y^2 + a^{-3} y^4).$$

Indeed, we have then

$$\ell_a(\varphi_F(x,a)) = \text{tr}_n\left(a^6(a^2 x + ax^2) + a^3(a^4 x^2 + a^2 x^4) + a^{-3}(a^8 x^4 + a^4 x^8)\right)$$
$$= \varphi_f(x,a)$$

and if there exists $x \in \mathbb{F}_2^n$ such that $\varphi_F(x,a) = 1$ then

$$\ell_a(1) = \text{tr}_n\left(a^{-3}\right) = \text{tr}_n\left(\frac{x}{a} + \left(\frac{x}{a}\right)^2\right) = 0.$$ □

Remark 18 [16] The APN property of the function $x^3 + \text{tr}_n(x^9)$ can be proven also with the following arguments due to Dillon [24]. If F is a quadratic function then for any nonzero a and for

$$\varphi_F(x,y) = F(x+y) + F(x) + F(y) + F(0)$$

there exists a linear function L_a such that

$$\varphi_F(ax,a) = L_a(x + x^2).$$

Indeed, if

$$F(x) = \sum_{i \le j} c_{i,j} x^{2^i + 2^j}$$

then

$$L_a(z) = \sum_{i \le j} c_{i,j} a^{2^i + 2^j} (T_{j-i}(z))^{2^i},$$

where

$$T_k(z) = z + z^2 + \dots + z^{2^{k-1}}.$$

Thus, F is APN if and only if for any nonzero a and z the equality $L_a(z) = 0$ implies $\mathrm{tr}_n(z) = 1$. In the case when $F(x) = x^3 + \mathrm{tr}_n(x^9)$ we have

$$L_a(z) = a^3 z + \mathrm{tr}_N(a^9 T_3(z))$$

and if $L_a(z) = 0$ for some $z \ne 0$ then $1 = a^3 z = \mathrm{tr}_n(a^9 T_3(z))$ which implies $1 = \mathrm{tr}_n(z^{-3}(z + z^2 + z^4)) = \mathrm{tr}_n(z)$. $\square$

Another class of APN functions, to which the construction of Theorem 31 can be applied, is a class of trinomial APN functions described in [11] (see case (3) in Table 2.6). However, for this class of functions we were able to construct only functions that are EA-equivalent to the original trinomial. More precisely we have the following proposition.

Proposition 25 [16] *Let m be a positive odd integer, $n = 2m$, α a primitive element of $\mathbb{F}_{2^n}$. Then the functions $F, G : \mathbb{F}_{2^n} \to \mathbb{F}_{2^n}$ with*

$$F(x) = x^6 + x^{2^m + 1} + \alpha^{2^m - 1} x^{6 \cdot 2^m},$$

$$G(x) = F(x) + \mathrm{tr}_n(\alpha^{2^{m-1} + 1} x^3),$$

are EA-equivalent.

Proof Let $t = \frac{\alpha^{2^{m+1} + 1}}{\alpha^{2^m - 1} + 1}$ and $L(x) = x + \mathrm{tr}_n(tx)$. It is not difficult to see that $L(F(x)) = G(x)$. $\square$

An Algorithmic Approach Below we describe an algorithmic approach to search for functions fulfilling the conditions of Theorem 31 when F is a Gold function. The first step will be to find an explicit description of the linear function l_a used in Theorem 31. Let $F(x) = x^{2^r + 1}$ and $f(x) = \mathrm{tr}_n(x^{2^i + 1})$. Then

$$\varphi_F(x, a) = a^{2^r + 1} \left(\left(\frac{x}{a} \right) + \left(\frac{x}{a} \right)^{2^r} \right)$$

and

$$\varphi_f(x, a) = \mathrm{tr}_n \left(a^{2^i + 1} \left(\left(\frac{x}{a} \right) + \left(\frac{x}{a} \right)^{2^i} \right) \right).$$

If we define $t = (ir^{-1} - 1) \bmod n$ we get

$$\varphi_f(x,a) = \mathrm{tr}_n\left(a^{2^i+1}\left(\left(\frac{x}{a}\right) + \left(\frac{x}{a}\right)^{2^i}\right)\right)$$

$$= \mathrm{tr}_n\left(a^{2^i+1}\left(\sum_{j=0}^{t}\left[\left(\frac{x}{a}\right) + \left(\frac{x}{a}\right)^{2^r}\right]^{2^{jr}}\right)\right)$$

$$= \mathrm{tr}_n\left(a^{2^i+1}\left(\sum_{j=0}^{t}\left[\frac{\varphi_F(x,a)}{a^{2^r+1}}\right]^{2^{jr}}\right)\right)$$

$$= \mathrm{tr}_n\left(\sum_{j=0}^{t} a^{2^i+1-(2^r+1)2^{jr}}\varphi_F(x,a)^{2^j}\right)$$

$$= \mathrm{tr}_n\left(\left(\sum_{j=0}^{t} a^{2^{i-jr}+2^{-jr}-(2^r+1)}\right)\varphi_F(x)\right).$$

Thus denoting

$$T_i^r(a) = \sum_{j=0}^{t} a^{2^{i-jr}+2^{-jr}-(2^r+1)}$$

we get

$$\varphi_f(x,a) = \mathrm{tr}_n(T_i^r(a)\varphi_F(x,a)).$$

In general for $g(x) = \sum_i \alpha_i x^{2^i+1}$ we get

$$\varphi_g(x,a) = \mathrm{tr}_n\left(\left(\sum_i \alpha_i T_i^r(a)\right)\varphi_F(x,a)\right).$$

Following Theorem 31, the condition for $F + g$ to be APN is that, if $\mathrm{tr}_n(a^{-(2^r+1)}) = 0$ then

$$\mathrm{tr}_n\left(\sum_i \alpha_i T_i(a)\right) = \sum_i \mathrm{tr}_n\left(\alpha_i T_i^r(a)\right) = 0.$$

Fixing a base $(b_j)_j$ of $\mathbb{F}_{2^n}$ over $\mathbb{F}_2$ we can consider the set of vectors

$$\left\{\mathrm{tr}_n(b_j T_i^r(a))_{a \in \mathbb{F}_{2^n}, \mathrm{tr}_n(a^{-3})=0} \mid i, j \in \{0 \ldots n-1\}\right\}.$$

Given F, finding a quadratic function g such that the conditions of Theorem 31 are fulfilled is equivalent to finding a set of linearly dependent vectors in this set. We computed these vectors and all linear dependent sets up to dimension 15. The only examples in addition to $x^3 + \mathrm{tr}_n(x^9)$ are listed below.

1. If n is even, then the function $\mathrm{tr}_n \circ T^r_{n/2}$ is constant zero. Thus in this case we can always add $\mathrm{tr}_n(x^{2^{n/2}+1})$. However this function is constant zero.
2. For $n = 5$ the function $x^5 + \mathrm{tr}_n(x^3)$ is APN.
3. For $n = 8$ the function $x^9 + \mathrm{tr}_n(x^3)$ is APN.

CCZ-Inequivalence of the New APN Function to Power Mappings
Theorem 32 [16] *The function of Corollary 25 is CCZ-inequivalent to any Gold function on $\mathbb{F}_{2^n}$ if $n \geq 7$ and $n > 2p$ where p is the smallest positive integer different from 1 and 3 and coprime with n.*

Proof Let $F(x) = x^3 + \mathrm{tr}_n(x^9)$ and $G(x) = x^{2^r+1}$ be APN functions on $\mathbb{F}_{2^n}$, $n \geq 7$, $r \leq (n-1)/2$.

Suppose the functions F and G are EA-equivalent. Then, there exist affine permutations L_1, L_2 and an affine function L' such that

$$L_1(x^3) + L_1(\mathrm{tr}_n(x^9)) = (L_2(x))^{2^t+1} + L'(x).$$

That is,

$$L_1(x^3) + L_1(1)\mathrm{tr}_n(x^9) = (L_2(x))^{2^t+1} + L'(x).$$

Since the functions are quadratic, we can assume without loss of generality that L_1 and L_2 are linear: $L_1(x) = \sum_{m \in \mathbb{Z}/n\mathbb{Z}} b_m x^{2^m}$, $L_2(x) = \sum_{p \in \mathbb{Z}/n\mathbb{Z}} c_p x^{2^p}$. Then we get

$$\sum_{m \in \mathbb{Z}/n\mathbb{Z}} b_m x^{3 \cdot 2^m} + \mathrm{tr}_n(x^9) \sum_{m \in \mathbb{Z}/n\mathbb{Z}} b_m = \sum_{l,p \in \mathbb{Z}/n\mathbb{Z}} c_p c_l^{2^t} x^{2^{l+t}+2^p} + L'(x). \qquad (5.30)$$

On the left hand side of the identity (5.30) we have only items of the type $x^{3 \cdot 2^m}$, $x^{9 \cdot 2^m}$, with some coefficients. Therefore this must be true also for the right hand side of the identity.

Let p be the smallest positive integer different from 1 and 3 such that $\gcd(n, p) = 1$ (for example, if n is odd then $p = 2$, if n is even and not divisible by 5 then $p = 5$). If $n > 2p$ then $2^p + 1$ is not in the same cyclotomic coset with 3 or 9. Therefore, the items of the type $x^{2^k(2^p+1)}$ must cancel. That is, for any k

$$c_k c_{k-t+p}^{2^t} = c_{k+p} c_{k-t}^{2^t}. \qquad (5.31)$$

Since $n \geq 7$ then 3 and 9 are in different cyclotomic cosets and we have for any k

$$L_1(1) = c_k c_{k-t+3}^{2^t} + c_{k+3} c_{k-t}^{2^t}.$$

If $L_1(1) \neq 0$ then

$$c_k c_{k-t+3}^{2^t} \neq c_{k+3} c_{k-t}^{2^t}. \qquad (5.32)$$

If $c_k \neq 0$ for all k then from (5.31) and (5.32) we get

$$c_k c_{k-t}^{-2^t} = c_{k+p} c_{k-t+p}^{-2^t}, \tag{5.33}$$

$$c_k c_{k-t}^{-2^t} \neq c_{k+3} c_{k-t+3}^{-2^t}. \tag{5.34}$$

Since $\gcd(n, p) = 1$ and from (5.33)

$$c_k c_{k-t}^{-2^t} = c_m c_{m-t}^{-2^t}$$

for any m. It contradicts (5.34). Thus, $c_k = 0$ for some k. Then from (5.31) and (5.32) we get that $c_{k+p} = 0$. Repeating this step for $c_{k+p}, c_{k+2p}, \ldots$ we get $c_{k+ps} = 0$ and since $\gcd(n, p) = 1$ then $c_k = 0$ for all k. A contradiction. If $L_1(1) = 0$ then the equation $L(x) = 0$ has at least 2 solutions 0, 1 and therefore L_1 is not a permutation. Thus, F and G are EA-inequivalent.

Suppose that $F(x)$ and $G(x)$ are CCZ-equivalent, that is, there exists an affine automorphism $L = (L_1, L_2)$ of $\mathbb{F}_{2^n} \times \mathbb{F}_{2^n}$ such that $y = F(x) \Leftrightarrow L_2(x, y) = G(L_1(x, y))$ and $L_1(x, F(x))$ is a permutation. This implies then $L_2(x, F(x)) = G(L_1(x, F(x)))$. Writing $L_1(x, y) = L(x) + L'(y)$ and $L_2(x, y) = L''(x) + L'''(y)$ gives

$$L''(x) + L'''(F(x)) = G\left(L(x) + L'(F(x))\right). \tag{5.35}$$

We can write

$$L(x) = b + \sum_{m \in \mathbb{Z}/n\mathbb{Z}} b_m x^{2^m},$$

$$L'(x) = b' + \sum_{m \in \mathbb{Z}/n\mathbb{Z}} b'_m x^{2^m},$$

$$L''(x) = b'' + \sum_{m \in \mathbb{Z}/n\mathbb{Z}} b''_m x^{2^m},$$

$$L'''(x) = b''' + \sum_{m \in \mathbb{Z}/n\mathbb{Z}} b'''_m x^{2^m},$$

$$b + b' = c.$$

Then we get

$$G(L(x) + L'(F(x))) = \left(L(x) + L'(x^3 + \mathrm{tr}_n(x^9))\right)$$
$$\times \left(L(x) + L'(x^3 + \mathrm{tr}_n(x^9))\right)^{2^r}$$
$$= [\sum_{m,k \in \mathbb{Z}/n\mathbb{Z}} b_m b_k'^{2^r} x^{2^m + 2^{k+r} + 2^{k+r+1}}$$
$$+ L'(1)^{2^r} \sum_{m,k \in \mathbb{Z}/n\mathbb{Z}} b_m x^{2^m + 2^{k+3} + 2^k}$$

$$+ \sum_{m,k\in\mathbb{Z}/n\mathbb{Z}} b'_m b_k^{2^r} x^{2^{m+1}+2^m+2^{k+r}}$$

$$+ L'(1) \sum_{m,k\in\mathbb{Z}/n\mathbb{Z}} b_m^{2^r} x^{2^{m+r}+2^{k+3}+2^k)}]$$

$$+ [\sum_{m,k\in\mathbb{Z}/n\mathbb{Z}} b'_m b'^{2^r}_k x^{2^{m+1}+2^m+2^{k+r+1}+2^{k+r}}$$

$$+ L'(1)^{2^r} \sum_{m,k\in\mathbb{Z}/n\mathbb{Z}} b'_m x^{2^{m+1}+2^m+2^{k+3}+2^k}$$

$$+ L'(1) \sum_{m,k\in\mathbb{Z}/n\mathbb{Z}} b'^{2^r}_m x^{2^{m+r+1}+2^{m+r}+2^{k+3}+2^k}]$$

$$+ Q(x),$$

where $Q(x)$ is a quadratic polynomial. Obviously, all terms in the expression above whose exponents have 2-weight strictly greater than 2 must cancel.

Since F and G are EA-inequivalent then L' is not a constant. Then there exists $m \in \mathbb{Z}/n\mathbb{Z}$ such that $b'_m \neq 0$.

Let $L'(1) \neq 0$. Since the items with the exponent $2^{m+1} + 2^m + 2^{m+2} + 2^{m+5}$ have to vanish then we get

$$L'(1)^{2^r} b'_m = L'(1) b'^{2^r}_{m-r},$$

and since $L'(1) \neq 0, b'_m \neq 0$ and r is coprime with n then $b'_k \neq 0$ and

$$b'_k b'^{-2^r}_{k-r} = L'(1)^{1-2^r}$$

for all k. Now we can deduce that $b'_{k+r} = L'(1)^{1-2^r} b'^{2^r}_k$ for all k. Then, introducing μ such that

$$L'(1)^{1-2^r} = \mu^{2^r-1},$$

we deduce that

$$\mu b'_{k+r} = (\mu b'_k)^{2^r}$$

for all k and then that

$$\mu b'_{k+1} = (\mu b'_k)^2$$

(using that $\gcd(r,n) = 1$) and then

$$\mu b'_k = (\mu b'_0)^{2^k}.$$

This means that

$$\mu L'(x) = \mu b' + \mathrm{tr}_n(\mu b'_0 x).$$

It implies that all nonquadratic items in the last bracket vanish and $L'(x) = d + \text{tr}_n(d'x)$ for some d, d'.

The function L is not 0 because L' is not a permutation, then $b_m \neq 0$ for some m. Since the items with the exponent $2^m + 2^{m+2} + 2^{m+5}$ have to vanish then

$$L'(1)^{2^r} b_m = L'(1) b_{m-r}^{2^r}.$$

Like above we get $L(x) = d + \text{tr}_n(d'x)$. Thus,

$$L_1(x, F(x)) = d'' + \text{tr}_n(F'(x))$$

for some d'' and $F'(x)$ and $L_1(x, F(x))$ is not a permutation. A contradiction.

Let $L'(1) = 0$ and $r \neq 1$. Then $2^{m+1} + 2^m + 2^{m+r+1} + 2^{m+r}$ has 2-weight 4 and since the items with this exponent should cancel then we get

$$b'^{2^r+1}_m = b'_{m+r} b'^{2^r}_{m-r}.$$

Since $b'_m \neq 0$ then

$$b'_{m+r}, b'_{m-r} \neq 0$$

and

$$b'_m b'^{-2^r}_{m-r} = b'_{m+r} b'^{-2^r}_m.$$

Since $\gcd(n, r) = 1$ then $b'_k \neq 0$,

$$b'_k b'^{-2^r}_{k-r} = b'_m b'^{-2^r}_{m-r}$$

for all k and this implies $L'(x) = d + \text{tr}_n(d'x)$ for some d, d'. Since $L_1(x, F(x))$ is a permutation then $L \neq 0$ and $b_m \neq 0$ for some m. The items with the exponent $2^m + 2^{m+r} + 2^{m+r+1}$ should vanish. Therefore,

$$b_m b'^{2^r}_m = b'_{m+r} b^{2^r}_{m-r}$$

and

$$b_m b^{-2^r}_{m-r} = b'_{m+r} b'^{-2^r}_m.$$

As above it leads to the equality $L(x) = d + \text{tr}_n(d'x)$ which is in contradiction with $L_1(x, F(x))$ being a permutation.

Let $L'(1) = 0$ and $r = 1$. Since $L'(1) = 0$ and $b'_m \neq 0$ then there exists t such that $b'_{m+t} \neq 0$. If $t \neq -1, -2$ then $2^{m+1} + 2^m + 2^{m+t+2} + 2^{m+t+1}$ has 2-weight 4 and we get

$$b'_m b'^{2^r}_{m+t} = b'_{m+t+1} b'^{2^r}_{m-1}$$

and

$$b'_m b'^{-2^r}_{m-1} = b'_{m+t+1} b'^{-2^r}_{m+t}.$$

Therefore, $L'(x) = d + \text{tr}_n(d'x)$ for some d, d'. If $t \neq 1, 2$ then $2^{m+t+1} + 2^{m+t} + 2^{m+2} + 2^{m+1}$ has 2-weight 4 and we get

$$b'_{m+t}b'^{2^r}_m = b'_{m+1}b'^{2^r}_{m+t-1}$$

and again $L'(x) = d + \text{tr}_n(d'x)$ for some d, d'. Thus, $L \neq 0$ and then $b_m \neq 0$ for some m. Since the items with the exponent $2^m + 2^{m+2} + 2^{m+3}$ cancel then

$$b_m b'^{2^r}_{m+1} = b'_{m+2}b^{2^r}_{m-1}$$

and

$$b_m b^{-2^r}_{m-1} = b'_{m+2}b'^{-2^r}_{m+1}.$$

This implies $L(x) = d + \text{tr}_n(d'x)$ and, thus, $L_1(x, F(x))$ is not a permutation. Therefore, F and G are not CCZ-equivalent. $\square$

Corollary 26 [16] *The function of Corollary 25 is EA-inequivalent to any power function on $\mathbb{F}_{2^n}$ if $n \geq 7$ and $n > 2p$, where p is the smallest positive integer different from 1 and 3 and coprime with n.*

Proof The function $F(x) = x^3 + \text{tr}_n(x^9)$ is quadratic APN and by Theorem 32 it is EA-inequivalent to any quadratic power function. Since the algebraic degree is EA-invariant then F is EA-inequivalent to any power mapping. $\square$

Dobbertin and inverse APN functions have unique Walsh spectra (except the case $n = 3$ when the inverse function is EA-equivalent to x^3) which are different from the Walsh spectra of quadratic APN functions (see [17, 21, 31]). Since the extended Walsh spectrum of a function is invariant under CCZ-equivalence then the function of Corollary 25 is CCZ-inequivalent to the inverse and Dobbertin APN functions for $n \geq 7$.

For $n = 7$ the Δ-rank of the function $F(x) = x^3 + \text{tr}_n(x^9)$ equals 212 and differs from the Δ-ranks of the Kasami functions x^{13} and x^{23} (which equal 338 and 436, respectively). Thus, for $n = 7$ the function F is CCZ-inequivalent to Kasami functions, and by Theorem 32 to the Gold functions. Since in this field the Welch and Niho cases coincide with the Kasami cases then F is CCZ-inequivalent to all power maps on $\mathbb{F}_{2^7}$. Further we used coding approach to check CCZ-inequivalence for $8 \leq n \leq 10$.

Corollary 27 [16] *The function $F(x) = x^3 + \text{tr}_n(x^9)$ is CCZ-inequivalent to power functions on $\mathbb{F}_{2^n}$ for $7 \leq n \leq 10$.*

Applying CCZ-equivalence to the quadratic APN function $F(x) = x^3 + \text{tr}_n(x^9)$, it is possible to construct classes of *nonquadratic* APN mappings which are CCZ-inequivalent to power functions.

Proposition 26 [16] *Let $F : \mathbb{F}_{2^n} \to \mathbb{F}_{2^n}$, $F(x) = x^3 + \text{tr}_n(x^9)$ then the following functions are CCZ-equivalent to F*

1) *for n odd the function with algebraic degree* 3

$$x^3 + \mathrm{tr}_n(x^9) + (x^2 + x)\mathrm{tr}_n(x^3 + x^9);$$

2) *for n even the function with algebraic degree* 3

$$x^3 + \mathrm{tr}_n(x^9) + (x^2 + x + 1)\mathrm{tr}_n(x^3);$$

3) *for n divisible by 6 the function with algebraic degree* 4

$$\left(x + \mathrm{tr}_n^3(x^6 + x^{12}) + \mathrm{tr}_n(x)\mathrm{tr}_n^3(x^3 + x^{12})\right)^3$$
$$+ \mathrm{tr}_n\left(\left(x + \mathrm{tr}_n^3(x^6 + x^{12}) + \mathrm{tr}_n(x)\mathrm{tr}_n^3(x^3 + x^{12})\right)^9\right);$$

4) *for n odd and divisible by* 3 *the function with algebraic degree* 4

$$\left(x^{\frac{1}{3}} + \mathrm{tr}_n^3(x + x^4)\right)^{-1} + \mathrm{tr}_n\left(\left(\left(x^{\frac{1}{3}} + \mathrm{tr}_n^3(x + x^4)\right)^{-1}\right)^9\right).$$

Proof The proof is the same as for the cases from [10, 13] (use the affine permutation $\mathcal{L}(x, y) = (x + \mathrm{tr}_n(y), y)$ for the first two cases, $\mathcal{L}(x, y) = (x + \mathrm{tr}_n^3(y^2 + y^4), y)$ for the third case and $\mathcal{L}(x, y) = (x + \mathrm{tr}_n^3(y + y^4), y)$ for the fourth case). □

Remark 19 [16] Note that the second and the fourth APN functions in Proposition 26 can be obtained from respectively the first and the third functions of Table 2.6 by adding $\mathrm{tr}_n(G(x))$ for some G. It means that the construction $F(x) + \mathrm{tr}_n(G(x))$, where F is APN and G is arbitrary, actually gives new APN functions even if F and G are not quadratic. For example, the only known APN polynomial which is CCZ-inequivalent to both power functions and to quadratic functions is a sum of some quadratic APN function over $\mathbb{F}_{2^6}$ with $\mathrm{tr}_6(G(x))$ for some non-quadratic function G (see [27]). □

Further Quadratic APN Constructions? There is a straightforward generalization of Theorem 31:

Theorem 33 [16] *Let F be a quadratic APN function from $\mathbb{F}_{2^n}$ to itself, let f be a quadratic function from $\mathbb{F}_{2^n}$ to $\mathbb{F}_{2^m}$ where m is a divisor of n, and*

$$\varphi_F(x, a) = F(x) + F(x + a) + F(a) + F(0),$$
$$\varphi_f(x, a) = f(x) + f(x + a) + f(a) + f(0).$$

Then the function $F(x) + f(x)$ is APN if for every nonzero $a \in \mathbb{F}_{2^n}$ there exists a linear function ℓ_a from $\mathbb{F}_{2^n}$ to $\mathbb{F}_{2^m}$ which satisfies the conditions

1) $\varphi_f(x, a) = \ell_a(\varphi_F(x, a))$,
2) *for every $u \in \mathbb{F}_{2^m}^*$, if $\varphi_F(x, a) = u$ for some $x \in \mathbb{F}_{2^n}$ then $\ell_a(u) \neq u$.*

We could find an application of Theorem 33:

Corollary 28 [16] *Let $n = 2m$ where m is an even positive integer. Let us denote by $\mathrm{tr}_{n/m}$ the trace function from $\mathbb{F}_{2^n}$ to $\mathbb{F}_{2^m}$: $\mathrm{tr}_n^m(x) = x + x^{2^m}$. The functions*

$$F(x) = x^3 + \mathrm{tr}_n^m(x^{2^m+2}) = x^3 + x^{2^m+2} + x^{2^{m+1}+1}$$

and

$$F'(x) = x^3 + (\mathrm{tr}_n^m(x))^3$$

are APN.

But unfortunately, these functions are EA-equivalent to power functions. Indeed, let G be the Gold function $G(x) = x^{2^{m-1}+1}$. Let γ be any element of $\mathbb{F}_4 \setminus \mathbb{F}_2$ and L_1, L_2 be the linear mappings $L_1(x) = \gamma^2 x^{2^{m+1}} + \gamma\, x^2$, $L_2(x) = \gamma\, x^{2^m} + \gamma^2 x$. Then $G \circ L_1(x) = L_2 \circ F(x)$.

5.5 Two More Classes of APN and AB Polynomials

In this section we generalize the construction

$$x^3 + \mathrm{tr}_n(x^9)$$

to the form

$$F(x) = L_1(x^3) + L_2(x^9)$$

where L_1 and L_2 are linear functions from $\mathbb{F}_{2^n}$ to itself, and we study conditions on L_1 and L_2 such that F is APN. In particular, we prove that, if n is even and the function $L_1(x) + L_2(x^3)$ is a permutation of $\mathbb{F}_{2^n}$, then F is APN. We have an example of the fact that even slightly changing functions L_1 and L_2 can lead to new cases of APN functions: for $n = 6, 8$, and a a primitive element of $\mathbb{F}_{2^n}^*$, the APN functions $x^3 + \mathrm{tr}_n(x^9)$ and $ax^3 + \mathrm{tr}_n(a^3 x^9)$ are CCZ-inequivalent. Besides, we construct a few families of APN functions: let n be a positive integer and $a \in \mathbb{F}_{2^n}^*$; the function

$$F_1(x) = x^3 + a^{-1}\mathrm{tr}_n(a^3 x^9)$$

and for n divisible by 3 the functions

$$F_2(x) = x^3 + a^{-1}\mathrm{tr}_n^3(a^6 x^{18} + a^{12} x^{36}),$$

$$F_3(x) = x^3 + a^{-1}\mathrm{tr}_n^3(a^3 x^9 + a^6 x^{18})$$

are APN over $\mathbb{F}_{2^n}$. As mentioned above, the function F_1 is CCZ-inequivalent to $x^3 + \mathrm{tr}_n(x^9)$ when $n = 6, 8$ and a is a primitive element of $\mathbb{F}_{2^6}^*$, and, therefore, we can see that F_1 defines a new class of APN functions which includes the class of functions $x^3 + \mathrm{tr}_n(x^9)$. Besides, we show that for $n = 9$, the functions F_1, F_2 and F_3 are pairwise CCZ-inequivalent, and the functions F_2 and F_3 are CCZ-inequivalent to any function from previously known families of APN functions. Hence, F_1 generalizes the family of functions $x^3 + \mathrm{tr}_n(x^9)$ while F_2 and F_3 define two new families of APN functions.

The Case n Even There is a sufficient condition for a function in even number of variables of the form $x^3 + L(x^9)$ (where L is linear) to be APN, which explains many cases of APN functions of this form.

Proposition 27 [15] *Let n be an even positive integer and L a linear function from $\mathbb{F}_{2^n}$ to itself. If the function*

$$F(x) = x + L(x^3)$$

is a permutation over $\mathbb{F}_{2^n}$ then $F(x^3)$ is APN on $\mathbb{F}_{2^n}$.

Proof Since the function $F'(x) = F(x^3)$ is quadratic then it is APN if and only if, for any $a \in \mathbb{F}_{2^n}^*$, the only solutions of the linear homogeneous equation

$$F'(ax) + F'(ax + a) + F'(a) = 0 \tag{5.36}$$

are $0, 1$. We get the equation

$$a^3(x + x^2) + L(a^9(x^8 + x)) = 0$$

and denoting $y = x + x^2$, we obtain

$$a^3 y + L(a^9(y^4 + y^2 + y)) = 0$$

with $\mathrm{tr}_n(y) = 0$. Hence, denoting $u = a^3 y$ and

$$A_u = \left\{ L\left(u^3 \left(y + \frac{1}{y} + \frac{1}{y^2} \right) \right) : \ y \in \mathbb{F}_{2^n}^*, \ \mathrm{tr}_n(y) = 0 \right\},$$

we see that if the condition

$$u \notin A_u \text{ for every } u \in \mathbb{F}_{2^n}^* \tag{5.37}$$

is satisfied then F' is APN. Note that

$$A_u \subseteq \{ L\left(u^3 z \right) : \ z \in \mathbb{F}_{2^n}, \ \mathrm{tr}_n(z) = 0 \}.$$

Since F is a permutation then $F(ux) + F(ux + u) \neq 0$ for any x and any nonzero u, that is,

$$u + L(u^3(x^2 + x + 1)) \neq 0.$$

Thus for any nonzero u

$$u \notin \{ L(u^3 v) : v \in \mathbb{F}_{2^n}, \ \mathrm{tr}_n(v) = \mathrm{tr}_n(1) \}. \tag{5.38}$$

For n even $\mathrm{tr}_n(1) = 0$ and

$$u \notin A_u \subseteq \{ L(u^3 v) : v \in \mathbb{F}_{2^n}, \ \mathrm{tr}_n(v) = \mathrm{tr}_n(1) \}.$$

Hence, condition (5.37) is satisfied and F' is indeed APN. □

Remark 20 [15] *Note that condition (5.37) has to be checked only for the elements u in the image of L. For instance, in the case* $L(x) = tr_n(x)$, *this condition has to be checked for* $u = 1$ *only and it is obviously satisfied. This gives a simpler way of proving that* $x^3 + tr_n(x^9)$ *is APN for every n (it is also easy to show that* $x + tr_n(x^3)$ *is bijective).*

Thus, to construct APN functions using Proposition 27, one has to look for permutations of the form $x + L(x^3)$ for a linear function L. It happens that examples of such permutations have been already found [13], in the framework of APN functions but in a different context. Note, that if $x + L(x^3)$ is a permutation then $x + a^{-1}L(a^3x^3)$ is also a permutation for every $a \neq 0$. Up to this equivalence, there are three known classes of such permutations:

- $x + tr_n(x^3)$ for any even n.
- $x + tr_n^3(x^{12} + x^6)$ for n divisible by 6.
- $x + tr_n^3(x^3 + x^6)$ for n divisible by 6.

We used Proposition 31 given below to search for more permutations of the desired type with the help of a computer. Using Proposition 31 a quite efficient backtracking search can be implemented. This approach allows to search for all such permutations up to $n = 14$ within a few days. Surprisingly, it turns out that the three classes listed above already cover all the cases for $n < 16$, n even. We pose the following open problem.

Open Problem Find more permutations of the form $x + L(x^3)$ not listed above or prove that the list above is already complete.

Next, we apply Proposition 27 to the second and the third classes of permutations listed above to construct APN functions (the first class gives the known APN function $x^3 + tr_n(x^9)$).

Corollary 29 [15] *Let n be a positive integer divisible by 6. Then the following functions from* $\mathbb{F}_{2^n}$ *to itself are APN*

$$x^3 + tr_n^3(x^{18} + x^{36}), \tag{5.39}$$

$$x^3 + tr_n^3(x^9 + x^{18}). \tag{5.40}$$

We extend now slightly Proposition 27. Let n and L, L' be linear functions from $\mathbb{F}_{2^n}$ to itself. According to Remark 3 of [13], if the function $F(x) = L'(x) + L(x^3)$ is a permutation over $\mathbb{F}_{2^n}$ then L' is a permutation. Hence, by Proposition 27, if F is a permutation then $F(x^3) = L'(x^3) + L(x^9)$ is APN. This fact is formulated in the following proposition.

Proposition 28 [15] *Let n be even and L and L' be linear functions from* $\mathbb{F}_{2^n}$ *to itself. If* $F(x) = L'(x) + L(x^3)$ *is a permutation over* $\mathbb{F}_{2^n}$ *then* $F(x^3)$ *is APN on* $\mathbb{F}_{2^n}$.

For instance, for n even, let $l(x) = ax + b$ with $a \in \mathbb{F}_{2^n}^*$, $b \in \mathbb{F}_{2^n}$, and let L be a linear function from $\mathbb{F}_{2^n}$ to itself. If $F(x) = x + L(x^3)$ is a permutation then

$$F(l(x)) = l(x) + L((l(x))^3)$$

$$= ax + b + L(a^3x^3 + a^2bx^2 + ab^2x + b^3)$$

$$= ax + L(a^2bx^2 + ab^2x) + L(a^3x^3) + b + L(b^3)$$

is a permutation too, and we can apply Proposition 28. We deduce:

Corollary 30 [15] *Let n be even, let $a \in \mathbb{F}_{2^n}^*$, $b \in \mathbb{F}_{2^n}$, and let L be a linear function from $\mathbb{F}_{2^n}$ to itself. If $x + L(x^3)$ is a permutation over $\mathbb{F}_{2^n}$ then the function*

$$ax^3 + L(a^3x^9 + a^2bx^6 + ab^2x^3)$$

is APN on $\mathbb{F}_{2^n}$.

Surprisingly enough, this almost straightforward corollary of Proposition 27 gives new examples of APN functions for different choices of elements a and b as we see it below.

Corollary 31 [15] *Let $a \in \mathbb{F}_{2^n}^*$. For n even the function*

$$x^3 + a^{-1}\mathrm{tr}_n(a^3x^9) \tag{5.41}$$

and for n divisible by 6 the functions

$$x^3 + a^{-1}\mathrm{tr}_n^3(a^6x^{18} + a^{12}x^{36}) \tag{5.42}$$

$$x^3 + a^{-1}\mathrm{tr}_n^3(a^3x^9 + a^6x^{18}) \tag{5.43}$$

are APN over $\mathbb{F}_{2^n}$.

Let α be a primitive element of $\mathbb{F}_{2^n}^*$ and $a = \alpha^{3i+j}$ for some i and $0 \le j \le 2$. Then

$$x^3 + a^{-1}\mathrm{tr}_n(a^3x^9) = a^{-1}(\alpha^j y^3 + \mathrm{tr}_n(\alpha^{3j}y^9))$$

where $y = \alpha^i x$. If $j = 2$ then taking $y = z^2$ we get

$$x^3 + a^{-1}\mathrm{tr}_n(a^3x^9) = a^{-1}(\alpha z^3 + \mathrm{tr}_n(\alpha^3 z^9))^2.$$

Hence, (5.41) can give at most two different cases $a = 1$ and $a = \alpha$ (since all other cases are EA-equivalent to these two as shown above). This is also true for functions (5.42) and (5.43), that is, each of them can give at most two different cases $a = 1$ and $a = \alpha$. An important observation is that for $a = 1$ and $a = \alpha$ those functions are not necessarily CCZ-equivalent. For example, consider $n = 6, 8$, then the APN function $x^3 + \mathrm{tr}_n(x^9)$ is CCZ-inequivalent to $x^3 + a^{-1}\mathrm{tr}_n(\alpha^3 x^9)$. Hence, function (5.41) defines a new class of APN functions which includes the class of functions $x^3 + \mathrm{tr}_n(x^9)$ from [16].

The Case n Odd For n even the function $x + \mathrm{tr}_n(x^3)$ and, when n is also divisible by 3, the functions $x + \mathrm{tr}_n^3(x^3 + x^6)$ and $x + \mathrm{tr}_n^3(x^6 + x^{12})$ are permutations while they are not permutations for n odd. However, as we shall see further, the corresponding functions $x^3 + \mathrm{tr}_n(x^9)$, $x^3 + \mathrm{tr}_n^3(x^9 + x^{18})$ and $x^3 + \mathrm{tr}_n^3(x^{18} + x^{36})$ are APN for both n even and n odd cases.

Besides, Proposition 27 does not apply to the case n odd. However, it can be adapted to this case. Assume first that $F(x) = x + L(x^3)$ is a permutation over the super-field $\mathbb{F}_{2^{2n}}$, then $F(x^3)$ is APN over $\mathbb{F}_{2^n}$, since we know then that, for every nonzero element u of $\mathbb{F}_{2^n}$, we have $u \notin \{L(u^3v), v \in \mathbb{F}_{2^{2n}}, \mathrm{tr}_{2n}(v) = 0\}$ and that these v include all the elements of $\mathbb{F}_{2^n}$ and a fortiori those of null trace tr_n. In fact, thanks to the observations in the proof of Proposition 27, this condition can be weakened and generalized:

Proposition 29 [15] *Let n be any positive integer and K some field extension of $\mathbb{F}_{2^n}$. Let L be an $\mathbb{F}_2$-linear mapping from $\mathbb{F}_{2^n}$ to $\mathbb{F}_{2^n}$ extended to an $\mathbb{F}_2$-linear mapping from K to K. Let E be a coset in K of a vector space containing $L(\mathbb{F}_{2^n})$. Assume that $F(x) = x + L(x^3)$ is injective on E and that the set $\{x^2 + x + 1; x \in E\}$ contains the set of elements y of $\mathbb{F}_{2^n}$ such that $\mathrm{tr}_n(y) = 0$. Then $F(x^3)$ is APN over $\mathbb{F}_{2^n}$.*

We can take $K = \mathbb{F}_{2^{2n}}$ and $E = w + \mathbb{F}_{2^n}$ where $w \in \mathbb{F}_4 \setminus \mathbb{F}_2$. If $F(x) = x + L(x^3)$ is injective on E then $F(x^3)$ is APN on $\mathbb{F}_{2^n}$ since the set

$$\{x^2 + x + 1 = x + w + (x + w)^2; x \in E\}$$

contains the set of elements y of $\mathbb{F}_{2^n}$ such that $\mathrm{tr}_n(y) = 0$.

For instance, if n is odd and divisible by 3 then the functions

$$F_1(x) = x + \mathrm{tr}_n^3(x^6 + x^{12}),$$

$$F_2(x) = x + \mathrm{tr}_n^3(x^3 + x^6)$$

are injective over $w + \mathbb{F}_{2^n} \subset \mathbb{F}_{2^{2n}}$, where $w \in \mathbb{F}_4 \setminus \mathbb{F}_2$. Indeed, the function F_2 is injective over $w + \mathbb{F}_{2^n}$ if and only if the function

$$x + w + \mathrm{tr}_n^3((x + w)^3 + (x + w)^6)$$

is injective over $\mathbb{F}_{2^n}$. We have

$$x + w + \mathrm{tr}_n^3((x + w)^3 + (x + w)^6)$$

$$= x + w + \mathrm{tr}_n^3(x^3 + x^6) + \mathrm{tr}_n^3(w^2x + w^2x^4 + w^3 + w^6)$$

$$= x + w + \mathrm{tr}_n^3(x^3 + x^6) + \mathrm{tr}_n^3(w^2(x + x^4))$$

because $w^4 = w$ an $w^3 = w^6 = 1$. Assume that there exist $x, a \in \mathbb{F}_{2^n}$, $a \neq 0$ such that

$$x + \mathrm{tr}_n^3(x^3 + x^6) + \mathrm{tr}_n^3(w^2(x + x^4))$$

$$= (x + a) + \mathrm{tr}_n^3((x + a)^3 + (x + a)^6) + \mathrm{tr}_n^3(w^2((x + a) + (x + a)^4)).$$

Then we get

$$a + \mathrm{tr}_n^3(ax^2 + a^2x + a^2x^4 + a^4x^2 + a^3 + a^6) + \mathrm{tr}_n^3(w^2(a + a^4)) = 0. \quad (5.44)$$

Note that since n is odd and $a \in \mathbb{F}_{2^n}$ then

$$(\mathrm{tr}_n^3(w^2(a + a^4)))^8 = \mathrm{tr}_n^3(w(a + a^4)^8) = \mathrm{tr}_n^3(w^{2^{n-3}}(a + a^4)) = \mathrm{tr}_n^3(w(a + a^4)).$$

Then adding (5.44) to its 8-th power we obtain

$$a + a^8 + \text{tr}_n^3((w^2 + w)(a + a^4)) = 0$$

which is actually

$$a + a^8 + \text{tr}_n^3(a + a^4) = 0$$

since $w^2 + w = 1$. Thus, $a + a^8 \in \mathbb{F}_8$ and

$$a + a^8 = (a + a^8)^8 = a^8 + a^{64},$$

which gives $a = a^{64}$. Hence, $a \in \mathbb{F}_{2^6} \cap \mathbb{F}_{2^n}^*$, and since n is odd then $a \in \mathbb{F}_{2^3}^*$ and

$$a + a^3 + a^6 + \text{tr}_n^3(a^2x + a^2x^4 + (a + a^4)x^2) + (a + a^4)\text{tr}_n^3(w^2) = 0.$$

Since $\text{tr}_n^3(w^2) = w^i$, where i equals either 1 or 2, then the previous equality gives

$$a + a^3 + a^6 + \text{tr}_n^3(a^2x + a^2x^4 + (a + a^4)x^2) = (a + a^4)w^i.$$

The left side of the equality above is in $\mathbb{F}_8$, the right in $w^i \cdot \mathbb{F}_8$. Hence, $a + a^4 = 0$ and $a \in \mathbb{F}_4 \cap \mathbb{F}_{2^3}^*$, and, therefore, $a = 1$. Then

$$0 = a + a^3 + a^6 + \text{tr}_n^3(a^2x + a^2x^4 + (a + a^4)x^2) = 1 + \text{tr}_n^3(x + x^4).$$

The equation $1 + \text{tr}_n^3(x + x^4) = 0$ has no solutions in $\mathbb{F}_{2^n}$. Indeed, if

$$\text{tr}_n^3(x + x^4) = 1 \tag{5.45}$$

then

$$\text{tr}_n^3(x + x^2) = (\text{tr}_n^3(x + x^4))^2 = 1, \tag{5.46}$$

and adding (5.45) and (5.46)

$$0 = \text{tr}_n^3(x + x^4 + x + x^2) = \text{tr}_n^3(x^2 + x^4) = (\text{tr}_n^3(x + x^2))^2 = 1.$$

A contradiction.

Thus, the function F_2 is indeed injective over $w + \mathbb{F}_{2^n} \subset \mathbb{F}_{2^{2n}}$, and the proof for the case of the function F_1 is similar. Therefore, by Proposition 29 the functions $F_1(x^3)$ and $F_2(x^3)$ are APN over $\mathbb{F}_{2^n}$ for n odd as well. Now we can formulate the following result.

Corollary 32 [15] *Let n be a positive integer and $a \in \mathbb{F}_{2^n}^*$. Then function (5.41) is APN over $\mathbb{F}_{2^n}$, and, if n is divisible by 3, functions (5.42) and (5.43) are APN over $\mathbb{F}_{2^n}$.*

Unlike to the case n even, for n odd different choices of $a \in \mathbb{F}_{2^n}^*$ for function (5.41) (respectively, functions (5.42) and (5.43)) give functions EA-equivalent to the case $a = 1$. Indeed, $\gcd(n, 3) = 1$ for n odd and, for example, $ax^3 + \text{tr}_n(a^3x^9) = y^3 + \text{tr}_n(y^9)$ with $y = a^{\frac{1}{3}}x$.

We checked with a computer (by testing the isomorphy of the associated binary codes, see [21]) that, for $n = 9$, functions (5.41), (5.42) and (5.43) are pairwise CCZ-inequivalent, and, that (5.42) and (5.43) are CCZ-inequivalent to any function from previously known families of APN functions. Hence, functions (5.42) and (5.43) define two new families of APN functions.

Sufficient Conditions for Permutations $x + L(x^3)$ To construct new APN functions thanks to Proposition 27, we need to find more linear mappings L such that $x + L(x^3)$ is a permutation. There is a simple sufficient condition for that.

Proposition 30 [15] *Let n be a positive integer and L be a linear function from $\mathbb{F}_{2^n}$ to itself. If for every $u \in \mathbb{F}_{2^n}$ such that $L(u) \neq 0$, the condition*

$$\mathrm{tr}_n \left(\frac{u}{(L(u))^3} \right) = \begin{cases} 0 \; \textit{if n is odd} \\ 1 \; \textit{if n is even} \end{cases}$$

is satisfied, then $F(x) = x + L(x^3)$ is a permutation.

Proof   Let $x, y \in \mathbb{F}_{2^n}$ be any elements such that

$$y = x + L(x^3). \tag{5.47}$$

Then, denoting $L^i(x) = (L(x))^i$, $i = 2, 3$, we have

$$x^3 = (y + L(x^3))^3 = y^3 + y^2 L(x^3) + y L^2(x^3) + L^3(x^3).$$

This means that there exists $z \in \mathbb{F}_{2^m}$ such that:

$$z = (y + L(z))^3 = y^3 + y^2 L(z) + y L^2(z) + L^3(z). \tag{5.48}$$

Then, F is a permutation if and only if, for every y, Equation (5.48) has at most one solution z. This is clear if n is odd since the function $x \to x^3$ being then bijective, relation (5.48) is equivalent to (5.47) with $z = x^3$. If n is even, then if for some y, there exist several z satisfying (5.48), then for each of them, $x = y + L(z)$ has z for cube and is therefore a solution of equation $y = x + L(x^3)$, which means that F is not injective; conversely, if for some y there exist several x satisfying (5.47), these elements cannot have same cube because of (5.47) and there are several z satisfying (5.48).

A sufficient condition for equation (5.48) to have at most one solution for each y is that, for every $u \neq 0$, there does not exist y and z such that

$$u = y^2 L(u) + y L^2(u) + L(z) L^2(u) + L^2(z) L(u) + L^3(u),$$

since if z and $z + u$ are solutions of (5.48), then this equality is satisfied. Note that $L(u)$ cannot be null since $u \neq 0$. We deduce that if, for every $u \neq 0$ such that $L(u) \neq 0$, there do not exist y and z such that

$$\frac{u}{L^3(u)} = \left(\frac{y}{L(u)} \right)^2 + \frac{y}{L(u)} + \left(\frac{L(z)}{L(u)} \right)^2 + \frac{L(z)}{L(u)} + 1$$

then F is a permutation. Hence a sufficient condition for F to be a permutation is $\mathrm{tr}_n \left(\frac{u}{L^3(u)} \right) = 1$, if n is even and $\mathrm{tr}_n \left(\frac{u}{L^3(u)} \right) = 0$, if n is odd. $\square$

We shall give now a necessary and sufficient condition for $x + L(x^3)$ being bijective which is in particular helpful for a computer search. This proposition is actually a special case of Proposition 6 of [13] for which we give a slightly different proof.

Proposition 31 [15] *Let n be an even integer and L be a linear function from $\mathbb{F}_{2^n}$ to itself. The function $x + L(x^3)$ is a permutation of $\mathbb{F}_{2^n}$ if and only if, for every $b \in \mathbb{F}_{2^n}^*$, such that $L^*(b) \neq 0$ there exists an element $\gamma \in \mathbb{F}_{2^n}$ such that $L^*(b) = \gamma^3$ and $\mathrm{tr}_n^2(\gamma^{-1}b) \neq 0$, were L^* denotes the adjoint linear mapping of L.*

Proof The mapping $F(x) = L(x^3) + x$ is a permutation if and only if, for every $b \neq 0$, the Boolean function $\mathrm{tr}_n(bF(x))$ is balanced. We calculate

$$\lambda_F(0, b) := \sum_x (-1)^{\mathrm{tr}_n(b(L(x^3)+x))}$$

$$= \sum_x (-1)^{\mathrm{tr}_n(L^*(b)x^3+bx)}$$

$$= \sum_x (-1)^{\mathrm{tr}_n(\frac{L^*(b)}{b^3}x^3+x)} = \lambda_{x^3}\left(1, \frac{L^*(b)}{b^3}\right).$$

Now we know that, for every $\beta \in \mathbb{F}_{2^n}$, the mapping $\mathrm{tr}_n(\beta x^3)$ is bent if and only if β is a non-cube (see [30]), and this implies that $\lambda_{x^3}(1, \beta)$ is not zero in this case. Thus a necessary condition for $L(x^3) + x$ to be a permutation is that $\frac{L^*(b)}{b^3}$ is a cube for every b. This is equivalent to $L^*(b)$ being a cube for every b. Denote by γ any third root of $L^*(b)$, i.e. $\gamma^3 = L^*(b)$. The next step is to actually compute the value $\lambda_{x^3}(1, \beta)$ in this case. To simplify notation we denote $L^*(b)/b^3$ by β. Then

$$\lambda_{x^3}(1, \beta)^2 = \sum_{x,y} (-1)^{\mathrm{tr}_n(\beta x^3+x+\beta y^3+y)} \qquad \bullet$$

$$= \sum_y (-1)^{\mathrm{tr}_n(\beta y^3+y)} \sum_x (-1)^{\mathrm{tr}_n(\beta(y^2x+yx^2))}$$

(by replacing y by $x + y$)

$$= \sum_y (-1)^{\mathrm{tr}_n(\beta y^3+y)} \sum_x (-1)^{\mathrm{tr}_n((\beta^2 y^4+\beta y)x^2)}.$$

Now the sum over x is 0 if $\beta^2 y^4 + \beta y \neq 0$ and 2^n otherwise. We denote $B = \gamma^{-1}b \cdot \mathbb{F}_4$. Since $(\gamma b^{-1})^3 = \beta$ then $\beta^2 y^4 + \beta y = 0$ if and only if $y \in B$. Then, using $\mathrm{tr}_n(1) = 0$, we get

$$\lambda_{x^3}(1, \beta)^2 = 2^n \sum_{y \in B} (-1)^{\mathrm{tr}_n(\beta y^3+y)} = 2^n \sum_{y \in B} (-1)^{\mathrm{tr}_n(y)}$$

$$= 2^n \sum_{y \in \mathbb{F}_4} (-1)^{\mathrm{tr}_n(\gamma^{-1}by)}$$

$$= 2^n \sum_{y \in \mathbb{F}_4} (-1)^{\mathrm{tr}_2(\mathrm{tr}_n^2(\gamma^{-1}b)y)}$$

$$= \begin{cases} 2^{n+2} & \text{if } \text{tr}_n^2(\gamma^{-1}b) = 0 \\ 0 & \text{else} \end{cases}.$$

This completes the proof.   □

A very similar characterization holds in the case n odd.

Proposition 32 [15] *Let n be an odd integer and L be a linear function from $\mathbb{F}_{2^n}$ to itself. The function $x + L(x^3)$ is a permutation of $\mathbb{F}_{2^n}$ if and only if for all $b \in \mathbb{F}_{2^n}$ either $L^*(b) = 0$ or $\text{tr}_n(\gamma^{-1}b) = 0$, were $L^*(b) = \gamma^3$ and L^* denotes the adjoint linear mapping of L.*

Proof Again, the mapping $F(x) = L(x^3) + x$ is a permutation if and only if, for every $b \neq 0$, the Boolean function $\text{tr}_n(bF(x))$ is balanced. A similar calculation as in the proof of Proposition 31 yields

$$\lambda_F(0, b) = \sum_x (-1)^{\text{tr}_n(b(L(x^3)+x))}$$

$$= \sum_x (-1)^{\text{tr}_n((L^*(b)^{1/3}x)^3+bx)}$$

$$= \sum_x (-1)^{\text{tr}_n\left(x^3+\frac{b}{L^*(b)^{1/3}}x\right)}$$

$$= \lambda_{x^3}\left(\frac{b}{L^*(b)^{1/3}}, 1\right).$$

Now in the case n odd $\lambda_{x^3}(a, 1) = 0$ if and only if $\text{tr}_n(a) = 0$ (see for example [19]) and thus F is a permutation if and only if $\text{tr}_n\left(\frac{b}{L^*(b)^{1/3}}\right) = 0$ for all b such that $L^*(b) \neq 0$. This completes the proof.   □

A little more can be said in the case where all coefficients of L are in the subfield $\mathbb{F}_{2^{n/2}}$, as it is the case for all the examples we know. For this we first state a technical lemma.

Lemma 5 [15] *Let n be an even integer and L a linear function from $\mathbb{F}_{2^n}$ to itself with $L(x) = \sum_{i=0}^{n-1} \lambda_i x^{2^i}$ where all $\lambda_i \in \mathbb{F}_{2^{n/2}}$. Furthermore denote by L' the restriction of L to $\mathbb{F}_{2^{n/2}}$. It holds that $L'^*(x) = L^*(x)$ for all $x \in \mathbb{F}_{2^{n/2}}$, i.e. the restriction of the adjoint operator of L to $\mathbb{F}_{2^{n/2}}$ is the adjoint operator of the restriction of L to $\mathbb{F}_{2^{n/2}}$.*

Proof First note that when all λ_i are in the subfield, so are the coefficients of L^*. Now let $x, y \in \mathbb{F}_{2^{n/2}}$ be arbitrary and choose $z \in \mathbb{F}_{2^n}$ such that $y = z^{2^{n/2}} + z$. We compute

$$\text{tr}_{n/2}(xL'(y)) = \text{tr}_{n/2}(xL(y)) = \text{tr}_{n/2}(xL(z + z^{2^{n/2}}))$$

$$= \text{tr}_{n/2}(xL(z) + (xL(z))^{2^{n/2}})$$

$$= \text{tr}_n(xL(z)) = \text{tr}_n(L^*(x)z)$$

$$= \text{tr}_{n/2}(L^*(x)(z + z^{2^{n/2}}))$$

$$= \text{tr}_{n/2}(L^*(x)y)$$

As L'^* is the unique linear mapping fulfilling $\text{tr}_{n/2}(xL'(y)) = \text{tr}_{n/2}(L'^*(x)y)$ the lemma follows.   □

Proposition 33 [15] *Let n be an even integer and L be a linear function from $\mathbb{F}_{2^n}$ to itself such that the function $x + L(x^3)$ is a permutation. If all coefficients of L are in $\mathbb{F}_{2^{n/2}}$ then L fulfills $L(\mathbb{F}_{2^{n/2}}) = \{0\}$ and $L^2 = 0$.*

Proof In this case $F(\mathbb{F}_{2^{n/2}}) \subseteq \mathbb{F}_{2^{n/2}}$ and therefore F is also bijective on this subfield. We first prove that $L^*(b) = 0$ for all $b \in \mathbb{F}_{2^{n/2}}$ and therefore that $L(b) = 0$ for all $b \in \mathbb{F}_{2^{n/2}}$. We consider the two cases $n/2$ even and odd separately. If $n/2$ is even then applying Proposition 31 on $x + L(x^3)$ viewed as a permutation on $\mathbb{F}_{2^n}$ gives $L^*(b) = 0$ or $L^*(b) = \gamma^3$ and $\text{tr}_n^2(\gamma^{-1}b) \neq 0$. But for $b \in \mathbb{F}_{2^{n/2}}$ the second condition is impossible as when γ^3 is in $\mathbb{F}_{2^{n/2}}$ so is γ and therefore

$$\text{tr}_n^2(\gamma^{-1}b) = \text{tr}_{n/2}^2(\gamma^{-1}b\text{tr}_n^{n/2}(1)) = 0.$$

Thus we see that $L^*(b) = 0$ for all $b \in \mathbb{F}_{2^{n/2}}$. In the case $n/2$ odd we apply Proposition 32 on $x + L(x^3)$ viewed as a permutation on $\mathbb{F}_{2^{n/2}}$ (and for $b \in \mathbb{F}_{2^{n/2}}$) and conclude that either $L^*(b) = 0$ or $L^*(b) = \gamma^3$ where $\text{tr}_{n/2}(\gamma^{-1}b) = 0$. On the other hand applying Proposition 31 on $x + L(x^3)$ viewed as a permutation on $\mathbb{F}_{2^n}$ and Lemma 5 yields in the case where $L^*(b) \neq 0$ that $\text{tr}_n^2(\gamma^{-1}b) \neq 0$, a contradiction as for an element $a \in \mathbb{F}_{2^{n/2}}$ we have $\text{tr}_{n/2}(a) = \text{tr}_n^2(a)$. Thus for all even n it holds that $L^*(b) = 0$ for all $b \in \mathbb{F}_{2^{n/2}}$. This implies that $L(b) = 0$ for all $b \in \mathbb{F}_{2^{n/2}}$. Now, in both cases ($n$ even and n odd), we choose $x \in \mathbb{F}_{2^n}$ arbitrarily and compute

$$L(x) + L(x)^{2^{n/2}} = L(x) + L(x^{2^{n/2}}) = L(x + x^{2^{n/2}}) = 0$$

which is equivalent to $L(x) \in \mathbb{F}_{2^{n/2}}$. Hence, $L(L(x)) = 0$ for any $x \in \mathbb{F}_{2^n}$.   □
Finally, we recall a characterization of permutation of the form $F(x) = L(x^3) + L'(x)$ given in [13].

Proposition 34 (Prop. 5 of [13]) *Let L and L' be linear functions from $\mathbb{F}_{2^n}$ to itself. The mapping $F(x) = L(x^3) + L'(x)$ is a permutation over $\mathbb{F}_{2^n}$ if and only if, for every $u \in \mathbb{F}_{2^n}^*$ and every v such that $\text{tr}_n(v) = \text{tr}_n(1)$, the condition $L(u^3v) \neq L'(u)$ holds.*

References

1. T. Bending, D. Fon-Der-Flaass. Crooked functions, bent functions and distance-regular graphs. *Electron. J. Comb.*, 5 (R34), 14, 1998.
2. J. Bierbrauer. New semifields, PN and APN functions. *Designs, Codes and Cryptography*, v. 54, pp. 189–200, 2010.

3. A. W. Bluher. On existence of Budaghyan-Carlet APN hexanomials. http://arxiv.org/abs/1208.2346 (2012)
4. C. Bracken, Z. Zha. On the Fourier Spectra of the Infinite Families of Quadratic APN Functions. *Finite Fields and Their Applications* 18(3), pp. 537–546, 2012.
5. C. Bracken, E. Byrne, N. Markin, G. McGuire. New families of quadratic almost perfect nonlinear trinomials and multinomials. Finite Fields and Their Applications **14**(3), pp. 703–714, 2008.
6. C. Bracken, E. Byrne, N. Markin, G. McGuire. On the Fourier spectrum of Binomial APN functions. *SIAM journal of Discrete Mathematics*, 23(2), pp. 596–608, 2009.
7. C. Bracken, E. Byrne, N. Markin, G. McGuire. A Few More Quadratic APN Functions. *Cryptography and Communications* 3(1), pp. 43–53, 2011.
8. C. Bracken, C. H. Tan, Y. Tan. Binomial differentially 4 uniform permutations with high nonlinearity. *Finite Fields and Their Applications* 18(3), pp. 537–546, 2012.
9. K. A. Browning, J. F. Dillon, R. E. Kibler, M. T. McQuistan. APN Polynomials and Related Codes. *Journal of Combinatorics, Information and System Science, Special Issue in honor of Prof. D.K Ray-Chaudhuri on the occasion of his 75th birthday*, vol. 34, no. 1–4, pp. 135–159, 2009.
10. L. Budaghyan. The Simplest Method for Constructing APN Polynomials EA-Inequivalent to Power Functions. *Proceedings of First International Workshop on Arithmetic of Finite Fields, WAIFI 2007, Lecture Notes in Computer Science* 4547, pp. 177–188, 2007.
11. L. Budaghyan and C. Carlet. Classes of Quadratic APN Trinomials and Hexanomials and Related Structures. *IEEE Trans. Inform. Theory*, vol. 54, no. 5, pp. 2354–2357, May 2008.
12. L. Budaghyan and A. Pott. On Differential Uniformity and Nonlinearity of Functions. Special Issue of *Discrete Mathematics devoted to "Combinatorics 2006"*, 309(2), pp. 371–384, 2009.
13. L. Budaghyan, C. Carlet, A. Pott. New Classes of Almost Bent and Almost Perfect Nonlinear Functions. *IEEE Trans. Inform. Theory*, vol. 52, no. 3, pp. 1141–1152, March 2006.
14. L. Budaghyan, C. Carlet, G. Leander. Two classes of quadratic APN binomials inequivalent to power functions. *IEEE Trans. Inform. Theory*, 54(9), pp. 4218–4229, 2008.
15. L. Budaghyan, C. Carlet, G. Leander. On a construction of quadratic APN functions. *Proceedings of IEEE Information Theory Workshop, ITW'09*, pp. 374–378, Taormina, Sicily, Oct. 2009.
16. L. Budaghyan, C. Carlet, G. Leander. Constructing new APN functions from known ones. *Finite Fields and Their Applications*, v. 15, issue 2, pp. 150–159, April 2009.
17. A. Canteaut, P. Charpin, H. Dobbertin. Weight divisibility of cyclic codes, highly nonlinear functions on $\mathbb{F}_{2^m}$, and crosscorrelation of maximum-length sequences. *SIAM Journal on Discrete Mathematics*, 13(1), pp. 105–138, 2000.
18. A. Canteaut, P. Charpin and H. Dobbertin. Binary m-sequences with three-valued crosscorrelation: A proof of Welch's conjecture. *IEEE Trans. Inform. Theory*, 46 (1), pp. 4–8, 2000.
19. C. Carlet. Vectorial Boolean Functions for Cryptography. Chapter of the monography *Boolean Methods and Models*, Yves Crama and Peter Hammer eds, Cambridge University Press, pp. 398–469, 2010.
20. C. Carlet. Relating three nonlinearity parameters of vectorial functions and building APN functions from bent functions. *Designs, Codes and Cryptography*, v. 59(1–3), pp. 89–109, 2011.
21. C. Carlet, P. Charpin and V. Zinoviev. Codes, bent functions and permutations suitable for DES-like cryptosystems. *Designs, Codes and Cryptography*, 15(2), pp. 125–156, 1998.
22. J. F. Dillon. Elementary Hadamard Difference sets. Ph. D. Thesis, Univ. of Maryland, 1974.
23. J. F. Dillon. APN Polynomials and Related Codes. *Polynomials over Finite Fields and Applications*, Banff International Research Station, Nov. 2006.
24. J. F. Dillon. Private communication, Feb. 2007.
25. H. Dobbertin. Almost perfect nonlinear power functions over $GF(2^n)$: a new case for n divisible by 5. *Proceedings of Finite Fields and Applications FQ5*, pp. 113–121, 2000.

26. H. Dobbertin, Uniformly representable permutation polynomials, T. Helleseth, P.V. Kumar and K. Yang eds., *Proceedings of "Sequences and their applications–SETA '01"*, Springer Verlag, London, 1–22, 2002.

27. Y. Edel and A. Pott. A new almost perfect nonlinear function which is not quadratic. *Advances in Mathematics of Communications* 3, no. 1, pp. 59–81, 2009.

28. Y. Edel, G. Kyureghyan and A. Pott. A new APN function which is not equivalent to a power mapping. *IEEE Trans. Inform. Theory*, vol. 52, no. 2, pp. 744–747, Feb. 2006.

29. G. Lachaud and J. Wolfmann. The Weights of the Orthogonals of the Extended Quadratic Binary Goppa Codes. *IEEE Trans. Inform. Theory*, vol. 36, pp. 686–692, 1990.

30. G. Leander. Monomial bent functions. *IEEE Transactions on Information Theory*, vol. 52, no. 2, pp. 738–743, 2006.

31. K. Nyberg. S-boxes and Round Functions with Controllable Linearity and Differential Uniformity. *Proceedings of Fast Software Encryption 1994, LNCS* 1008, pp. 111–130, 1995.

32. S. Yoshiara. Equivalence of quadratic APN functions. *J. Algebr. Comb.* 35, pp. 461–475, 2012.

Chapter 6
Construction of Planar Functions

6.1 On the Structure of This Chapter

In this chapter we present two infinite families of perfect nonlinear Dembowski-Ostrom multinomials over $\mathbb{F}_{p^{2k}}$ where p is any odd prime and k a positive integer. These are the families corresponding to case (iv) in Sect. 2.4.1 (where known families of PN functions are listed) which were first introduced in [3, 5]. We prove that in general these functions are CCZ-inequivalent to previously known PN mappings. One of these families has been constructed by extension of a known family of APN functions over $\mathbb{F}_{2^{2k}}$. This shows that known classes of APN functions over fields of even characteristic can serve as a source for further constructions of PN mappings over fields of odd characteristics. This method, first introduced in [3] was further applied to the families of APN binomials (1-2) from [6] to extend them to the families of planar binomials (v) and (vi).

Besides, we supply results indicating that planar functions (iv) define new commutative semifields. After the works of Dickson (1906) and Albert (1952), these were the firstly found infinite families of commutative semifields which are defined for all odd primes p.

Further we extend the family of PN functions from [1, 8] to a larger (up to CCZ-equivalence) family of PN functions corresponding to (vii). This is done by using isotopisms of semifields (which are not strong) [4]. That is, extending the family of PN functions we still stay within the same family of commutative semifields (up to isotopic equivalence).

6.2 Families of Planar Multinomials (iv)

First Family of Case (iv) In [9] Ness gives a list of planar DO trinomials over $\mathbb{F}_{p^n}$ for $p \leq 7, n \leq 8$ which were found with a computer. Investigation of these functions has led us to the following family of planar DO polynomials.

© Springer International Publishing Switzerland 2014
L. Budaghyan, *Construction and Analysis of Cryptographic Functions*,
DOI 10.1007/978-3-319-12991-4_6

Theorem 34 [3, 5] *Let p be an odd prime, s and k positive integers such that* $\gcd(p^s + 1, p^k + 1) \neq \gcd(p^s + 1, (p^k + 1)/2)$ *and* $\gcd(k + s, 2k) = \gcd(k + s, k)$. *Let also* $n = 2k$, $b \in \mathbb{F}_{p^n}^*$, *and* $\sum_{i=0}^{k-1} c_i x^{p^i}$ *be a permutation over* $\mathbb{F}_{p^k}$ *with coefficients in* $\mathbb{F}_{p^k}$. *Then the function*

$$F(x) = (bx)^{p^s+1} - \left((bx)^{p^s+1}\right)^{p^k} + \sum_{i=0}^{k-1} c_i x^{p^i(p^k+1)}$$

is PN over $\mathbb{F}_{p^n}$.

Proof Since F is DO polynomial then it is PN if for any $a \in \mathbb{F}_{p^n}^*$ the equation $F(x + a) - F(x) - F(a) = 0$ has only 0 as a solution. We have

$$\begin{aligned}
\Delta(x) &= F(x + a) - F(x) - F(a) \\
&= b^{p^s+1}(ax^{p^s} + a^{p^s}x) - b^{p^k(p^s+1)}(a^{p^k}x^{p^{k+s}} + a^{p^{k+s}}x^{p^k}) \\
&\quad + \sum_{i=0}^{k-1} c_i(a^{p^i}x^{p^{k+i}} + a^{p^{k+i}}x^{p^i}).
\end{aligned}$$

Any solution of the equation $\Delta(x) = 0$ is also a solution of $\Delta(x) + \Delta(x)^{p^k} = 0$ and $\Delta(x) - \Delta(x)^{p^k} = 0$, that is, a solution of

$$\sum_{i=0}^{k-1} c_i(a^{p^i}x^{p^{k+i}} + a^{p^{k+i}}x^{p^i}) = 0, \tag{6.1}$$

$$b^{p^s+1}(ax^{p^s} + a^{p^s}x) = b^{p^k(p^s+1)}(a^{p^k}x^{p^{k+s}} + a^{p^{k+s}}x^{p^k}). \tag{6.2}$$

Since $\sum_{i=0}^{k-1} c_i x^{p^i}$ is a permutation over $\mathbb{F}_{p^k}$ then (6.1) implies

$$ax^{p^k} = -a^{p^k}x. \tag{6.3}$$

Now we can substitute ax^{p^k} in (6.2) by $-a^{p^k}x$ and then obtain

$$b^{p^s+1}(ax^{p^s} + a^{p^s}x) = -b^{p^k(p^s+1)}(a^{p^{k+s}+p^k-p^s}x^{p^s} + a^{p^{k+s}+p^k-1}x),$$

that is,

$$(b^{p^s+1}a + b^{p^k(p^s+1)}a^{p^{k+s}+p^k-p^s})x^{p^s} = -(b^{p^s+1}a^{p^s} + b^{p^k(p^s+1)}a^{p^{k+s}+p^k-1})x,$$

and since $a, b \neq 0$ then for $x \neq 0$

$$x^{p^s-1} = -\frac{b^{p^s+1}a^{p^s} + b^{p^k(p^s+1)}a^{p^{k+s}+p^k-1}}{b^{p^s+1}a + b^{p^k(p^s+1)}a^{p^{k+s}+p^k-p^s}} = -a^{p^s-1}, \tag{6.4}$$

when

$$b^{(p^k-1)(p^s+1)}a^{p^{k+s}+p^k-p^s-1} \neq -1. \tag{6.5}$$

Now assume that for some nonzero a inequality (6.5) is wrong, that is,

$$(ba)^{(p^k-1)(p^s+1)} = -1.$$

Then -1 is a power of $(p^k - 1)(p^s + 1)$ which is in contradiction with $\gcd(p^s + 1, p^k + 1) \neq \gcd(p^s + 1, (p^k + 1)/2)$ since -1 is a power of $(p^n - 1)/2$.

From (6.3) and (6.4) we get

$$y^{p^k-1} = y^{p^s-1} = -1, \tag{6.6}$$

where $y = x/a$. Since $n = 2k$ then the first equality in (6.6) implies $y^{p^{k+s}} = y$, that is, $y \in \mathbb{F}_{p^{k+s}}$. Thus, if $\gcd(k + s, 2k) = \gcd(k + s, k)$ then $y \in \mathbb{F}_{p^{\gcd(k+s,k)}}$ which contradicts the second equality in (6.6), that is, $y^{p^k-1} = 1 \neq -1$, for any $y \neq 0$. Therefore, the only solution of $\Delta(x) = 0$ is $x = 0$. □

Second Family of Case (iv) Below we show that one of the ways to construct PN mappings is to extend a known family of APN functions over $\mathbb{F}_{2^n}$ to a family of PN functions over $\mathbb{F}_{p^n}$ for odd primes p. We construct a class of PN functions by following the pattern of APN multinomials over $\mathbb{F}_{2^{2k}}$ presented in [2] and corresponding to case (11) in Table 2.6.

Theorem 35 [3, 5] *Let p be an odd prime, s and k positive integers, $n = 2k$, and $\gcd(k+s, n) = \gcd(k+s, k)$. If $b \in \mathbb{F}_{p^n}^*$ is not a square, $c \in \mathbb{F}_{p^n} \setminus \mathbb{F}_{p^k}$, and $r_i \in \mathbb{F}_{p^k}$, $0 \le i < k$, then the function*

$$F(x) = \mathrm{tr}_{2k}^k(bx^{p^s+1}) + cx^{p^k+1} + \sum_{i=1}^{k-1} r_i x^{p^{k+i}+p^i}$$

is PN over $\mathbb{F}_{p^n}$.

Proof We have to show that for any $a \in \mathbb{F}_{p^n}^*$ the equation $\Delta(x) = 0$ has only 0 as a solution when

$$\Delta(x) = F(x + a) - F(x) - F(a)$$

$$= \mathrm{tr}_{2k}^k\left(b(x^{p^s}a + xa^{p^s})\right) + c(x^{p^k}a + xa^{p^k}) + \sum_{i=1}^{k-1} r_i(x^{p^{k+i}}a^{p^i} + x^{p^i}a^{p^{k+i}}).$$

After replacing x by ax we get

$$\Delta_1(x) = \Delta(ax) = \mathrm{tr}_{2k}^k\left(ba^{p^s+1}(x^{p^s} + x)\right) + ca^{p^k+1}(x^{p^k} + x)$$

$$+ \sum_{i=1}^{k-1} r_i a^{p^{k+i}+p^i}(x^{p^{k+i}} + x^{p^i}).$$

Since $\Delta_1(x) = 0$ then $\Delta_1(x) - \Delta_1(x)^{p^k} = 0$, that is,

$$(ca^{p^k+1} - c^{p^k} a^{p^k+1})(x^{p^k} + x) = 0.$$

Thus,

$$(c - c^{p^k})a^{p^k+1}(x^{p^k} + x) = 0$$

and, therefore, $x^{p^k} = -x$ since $c \in \mathbb{F}_{p^{2k}} \setminus \mathbb{F}_{p^k}$.

Substituting $x^{p^k} = -x$ in $\Delta_1(x) = 0$ we obtain

$$\Delta_1(x) = ba^{p^s+1}(x^{p^s} + x) + b^{p^k} a^{p^{s+k}+p^k}(x^{p^{s+k}} + x^{p^k})$$

$$= (ba^{p^s+1} - b^{p^k} a^{p^{s+k}+p^k})(x^{p^s} + x).$$

Hence, if

$$ba^{p^s+1} \neq b^{p^k} a^{p^{s+k}+p^k} \tag{6.7}$$

then $x^{p^s} = -x$.

Assume that $ba^{p^s+1} = b^{p^k} a^{p^{s+k}+p^k}$ for some nonzero a. Then we get equalities

$$b^{p^k-1} = a^{p^s+1-p^{s+k}-p^k} = a^{-(p^s+1)(p^k-1)} = a^{(p^{k+s}-1)(p^k-1)}$$

which imply that b is a power of $\gcd(p^s + 1, p^k + 1)$ and of $\gcd(p^{s+k} - 1, p^k + 1)$. Thus, inequality (6.7) holds for any $a \neq 0$ if b is not a power of $\gcd(p^s + 1, p^k + 1)$ or a power of $\gcd(p^{s+k} - 1, p^k + 1)$. Since $\gcd(p^s + 1, p^k + 1)$ and $\gcd(p^{s+k} - 1, p^k + 1)$ are even then we cannot have inequality (6.7) for any nonzero b but we have this inequality, in particular, when b is not a square in $\mathbb{F}_{p^n}^*$.

Since $x^{p^k} = -x$ and $x^{p^s} = -x$ then $x^{p^k} = x^{p^s}$ and then by taking the p^k-th power we get $x^{p^{k+s}} = x$. Hence, if $\gcd(k + s, 2k) = \gcd(k + s, k)$ then $x \in \mathbb{F}_{p^{\gcd(k+s,k)}}$ and $x^{p^{\gcd(k+s,k)}} = x$. But $x^{p^k} = -x$, which implies $x = 0$. $\square$

6.3 Inequivalence of Families (iv) to Previously Known PN Functions

The planar functions (iv) are defined over $\mathbb{F}_{p^{2k}}$ for any odd prime p. Hence, when proving CCZ-inequivalence to the known PN functions we mainly concentrate our attention on the functions (i), (ii), and (iii), which were the only previously known PN functions defined for any odd prime p.

In the proposition below we show that any function which is CCZ-equivalent to x^2 should have some monomial of the form x^{2p^t} for some t, $0 \leq t < n$, in its polynomial representation.

Proposition 35 [3, 5] *Let p be an odd prime and n be a positive integer. Any function F of the form*

$$F(x) = \sum_{0 \le k < j < n} a_{kj} x^{p^k + p^j}$$

over $\mathbb{F}_{p^n}$ is CCZ-inequivalent to x^2.

Proof Since x^2 is a planar DO polynomial then CCZ-equivalence of F to x^2 implies the linear equivalence, that is, the existence of linear permutations L_1 and L_2 such that

$$(L_1(x))^2 + L_2(F(x)) = 0. \tag{6.8}$$

Let

$$L_1(x) = \sum_{i=0}^{n-1} u_i x^{p^i}, \tag{6.9}$$

$$L_2(x) = \sum_{i=0}^{n-1} v_i x^{p^i}. \tag{6.10}$$

Then equality (6.8) implies

$$0 = \left(\sum_{i=0}^{n-1} u_i x^{p^i} \right)^2 + \sum_{i=0}^{n-1} v_i \left(\sum_{0 \le k < j < n} a_{kj} x^{p^k + p^j} \right)^{p^i}$$

$$= \sum_{i=0}^{n-1} u_i^2 x^{2p^i} + 2 \sum_{0 \le i < j < n} u_i u_j x^{p^i + p^j} + \sum_{0 \le k < j < n, 0 \le i < n} v_i a_{kj}^{p^i} x^{p^i(p^k + p^j)}.$$

Since the identity above takes place for any $x \in \mathbb{F}_{p^n}$ then obviously $u_i^2 = 0$ for all $0 \le i < n$, that is, $L_1(x) = 0$. This contradicts the condition that L_1 is a permutation. Hence F is CCZ-inequivalent to x^2. □

Corollary 33 [3, 5] *The planar functions from the case (iv) are CCZ-inequivalent to x^2.* □

We give below a sufficient condition on DO polynomials to be CCZ-inequivalent to the PN functions of the case (ii).

Proposition 36 [3, 5] *Let p be an odd prime, n, n' and t positive integers such that $n' < n$ and $n/\gcd(n, t)$ is odd. Let a function $F : \mathbb{F}_{p^n} \to \mathbb{F}_{p^n}$ be such that*

$$F(x) = \sum_{i=0}^{n'} A_i(x^{p^{s_i}+1}),$$

where $0 < s_i < n$ and $s_i \ne s_j$, for all $i \ne j$, $0 \le i, j \le n'$, and the functions A_i, $0 \le i \le n'$, are linear. If $t \ne s_i$ and $t \ne n - s_i$ for all $0 \le i \le n'$ then the PN function $G(x) = x^{p^t+1}$ is CCZ-inequivalent to F.

Proof Assume that F and G are CCZ-equivalent. Since G is planar DO polynomial then CCZ-equivalence implies linear equivalence, that is, the existence of linear permutations L_1 and L_2, defined by (6.9–6.10), such that

$$G(L_1(x)) + L_2(F(x)) = 0.$$

We get

$$0 = \left(\sum_{i=0}^{n-1} u_i x^{p^i} \right)^{p^t+1} + \sum_{i=0}^{n-1} v_i \left(\sum_{i=0}^{n'} A_i (x^{p^{s_i}+1}) \right)^{p^i}$$

$$= \sum_{i,j=0}^{n-1} u_i u_j^{p^t} x^{p^i+p^{j+t}} + \sum_{i=0}^{n'} A_i'(x^{p^{s_i}+1}),$$

where A_i', $0 \le i \le n'$, are some linear functions. Since the latter expression is equal to 0 then the terms of the type x^{2p^i}, $0 \le i < n$, should vanish and we get

$$u_i u_{i-t}^{p^t} = 0, \qquad 0 \le i < n. \tag{6.11}$$

Since $t \ne s_i$ and $t \ne n - s_i$ for all $0 \le i \le n'$ then cancelling all terms of the type $x^{p^i(p^t+1)}$, $0 \le i < n$, we get

$$u_i u_i^{p^t} = -u_{i+t} u_{i-t}^{p^t}, \qquad 0 \le i < n. \tag{6.12}$$

Equalities (6.11) and (6.12) imply $L_1 = 0$. Indeed, if $u_i \ne 0$ for some i then from (6.11) we get $u_{i-t} = 0$ while from (6.12) we get $u_{i-t} \ne 0$. But L_1 is a permutation and cannot be constantly 0. This contradiction shows that the functions F and x^{p^t+1} are CCZ-inequivalent. $\qquad \square$

From proposition above we get the following straightforward corollaries.

Corollary 34 [3, 5] *The planar functions from the case (iv) are CCZ-inequivalent to* x^{p^t+1} *when* $s \ne \pm t$. $\qquad \square$

Corollary 35 [3, 5] *The planar functions from the case (iv) are CCZ-inequivalent to* x^{p^t+1} *when* $2k/\gcd(2k, s)$ *is even.* $\qquad \square$

Further we can prove that, under some conditions on coefficients, the first function of the case (iv) is CCZ-inequivalent also to x^{p^s+1}.

Proposition 37 [3, 5] *Let p be an odd prime, s and k positive integers, $n = 2k$, $n/\gcd(n, s)$ odd. The function*

$$F(x) = x^{p^s+1} - x^{p^{k+s}+p^s} \pm x^{p^k+1}$$

is CCZ-inequivalent to $G(x) = x^{p^s+1}$ *over* $\mathbb{F}_{p^n}$.

Proof Assume that F and G are CCZ-equivalent. Since G is planar DO polynomial then CCZ-equivalence implies the existence of linear permutations L_1 and L_2, defined by (6.9) and (6.10), such that

$$G(L_1(x)) + L_2(F(x)) = 0.$$

For $\epsilon \in \{1, -1\}$ we get

$$0 = \left(\sum_{i=0}^{n-1} u_i x^{p^i}\right)^{p^s+1} + \sum_{i=0}^{n-1} v_i \left(x^{p^s+1} - x^{p^{k+s}+p^s} + \epsilon x^{p^k+1}\right)^{p^i}$$

$$= \sum_{i,j=0}^{n-1} u_i u_j^{p^s} x^{p^i+p^{j+s}} + \sum_{i=0}^{n-1} v_i x^{p^{i+s}+p^i} - \sum_{i=0}^{n-1} v_i x^{p^{i+s+k}+p^{i+k}} + \epsilon \sum_{i=0}^{n-1} v_i x^{p^{i+k}+p^i}.$$

Since the latter expression is equal to 0 then the terms of the type x^{2p^i}, $0 \le i < n$, should vanish and we get

$$u_i u_{i-s}^{p^s} = 0, \qquad 0 \le i < n. \tag{6.13}$$

Considering items with exponents $p^{i+s}+p^i$ and with exponents $p^{i+k}+p^i$, $0 \le i < n$, we get

$$v_i - v_{i+k} + u_i u_i^{p^s} + u_{i+s} u_{i-s}^{p^s} = 0, \tag{6.14}$$

$$\epsilon v_i + u_i u_{i+k-s}^{p^s} + u_{i+k} u_{i-s}^{p^s} = 0. \tag{6.15}$$

Equality (6.15) implies

$$\epsilon v_i = -(u_i u_{i+k-s}^{p^s} + u_{i+k} u_{i-s}^{p^s}) = \epsilon v_{i+k}. \tag{6.16}$$

Equalities (6.14) and (6.16) imply

$$0 = v_i - v_{i+k} = -(u_i u_i^{p^s} + u_{i+s} u_{i-s}^{p^s}). \tag{6.17}$$

If $u_i \ne 0$ then $u_{i-s} = 0$ by (6.13). But if $u_{i-s} = 0$ then $u_i = 0$ by (6.17). Hence, $L_1 = 0$ which is impossible since L_1 is a permutation. This contradiction shows that the functions F and x^{p^s+1} are CCZ-inequivalent. $\square$

6.4 Nuclei of Semifields (iv)

It is proven in [7] that, for any planar DO function F, isotopism between the commutative semifield defined by F and a commutative twisted field, or the finite field, implies strong isotopism. Thus, PN functions of the case (iv) define commutative semifields nonisotopic to the field and to Albert's commutative twisted fields. Due to the theorem below we will see also that the commutative semifields corresponding to the first family of (iv) are also nonisotopic to Dickson semifields when k is odd and $b \in \mathbb{F}_{p^k}$.

Theorem 36 [5] *Let F be a PN function of the first family of the case (iv) with $b \in \mathbb{F}_{p^k}$. Then the middle nucleus of the commutative semifield defined by F has a square order.*

Proof For any $x, y \in \mathbb{F}_{p^{2k}}$ we denote

$$
\begin{aligned}
x \star y &= F(x + y) - F(x) - F(y) \\
&= b^{p^s+1}(xy^{p^s} + x^{p^s} y) - b^{p^k(p^s+1)}(x^{p^k} y^{p^{k+s}} + x^{p^{k+s}} y^{p^k}) \\
&\quad + \sum_{i=0}^{k-1} c_i (x^{p^i} y^{p^{k+i}} + x^{p^{k+i}} y^{p^i}),
\end{aligned}
\tag{6.18}
$$

and

$$
L(x) = 1 \star x = b^{p^s+1}\left(x + x^{p^s}\right) - b^{p^k(p^s+1)}\left(x^{p^k} + x^{p^{k+s}}\right) + \sum_{i=0}^{k-1} c_i \left(x^{p^i} + x^{p^{k+i}}\right).
$$

Then the multiplication $\circ$ of the commutative semifield $\mathbb{S}_F$ defined by F is

$$
x \circ y = L^{-1}(x) \star L^{-1}(y),
\tag{6.19}
$$

for any $x, y \in \mathbb{F}_{p^{2k}}$.

We are going to prove that for any $x, y \in \mathbb{F}_{p^{2k}}$ and any $\alpha \in \mathbb{F}_{p^2}$

$$
(x \circ L(\alpha)) \circ y = (y \circ L(\alpha)) \circ x,
$$

or, since L is a permutation

then, equivalently, we need to prove that

$$
(L(x) \circ L(\alpha)) \circ L(y) = (L(y) \circ L(\alpha)) \circ L(x),
$$

that is,

$$
L^{-1}(x \star \alpha) \star y = L^{-1}(y \star \alpha) \star x,
\tag{6.20}
$$

due to (6.19). We have

$$
L(x)^{p^k} + L(x) = 2 \sum_{i=0}^{k-1} c_i (x^{p^i} + x^{p^{k+i}}),
$$

$$
L(x)^{p^k} - L(x) = 2 b^{p^k(p^s+1)}(x^{p^k} + x^{p^{k+s}}) - 2 b^{p^s+1}(x + x^{p^s}).
$$

Note that $L(x^{p^k}) = L(x)^{p^k}$. Then applying L^{-1} to both sides of the equalities above we get

$$x^{p^k} + x = 2L^{-1}\Big(\sum_{i=0}^{k-1} c_i(x^{p^i} + x^{p^{k+i}})\Big), \tag{6.21}$$

$$x^{p^k} - x = 2L^{-1}\big(b^{p^k(p^s+1)}(x^{p^k} + x^{p^{k+s}}) - b^{p^s+1}(x + x^{p^s})\big). \tag{6.22}$$

Then, using (6.21–6.22) and $\alpha^{p^2} = \alpha$,

$$L^{-1}(x \star \alpha) = L^{-1}\big(b^{p^s+1}(x\alpha^{p^s} + x^{p^s}\alpha) - b^{p^k(p^s+1)}(x^{p^k}\alpha^{p^{k+s}} + x^{p^{k+s}}\alpha^{p^k})$$

$$+ \sum_{i=0}^{k-1} c_i(x^{p^i}\alpha^{p^{k+i}} + x^{p^{k+i}}\alpha^{p^i})\big)$$

$$= L^{-1}\big(b^{p^s+1}(x\alpha^{p^s} + (x\alpha^{p^s})^{p^s})$$

$$- b^{p^k(p^s+1)}((x\alpha^{p^s})^{p^k} + (x\alpha^{p^s})^{p^{k+s}}))$$

$$+ L^{-1}\big(\sum_{i=0}^{k-1} c_i((x\alpha^{p^k})^{p^i} + (x\alpha^{p^k})^{p^{k+i}})\big)$$

$$= -\frac{1}{2}((x\alpha^{p^s})^{p^k} - x\alpha^{p^s}) + \frac{1}{2}(x\alpha^{p^k} + (x\alpha^{p^k})^{p^k})$$

$$= \frac{1}{2}(\alpha^{p^s} + \alpha^{p^k})x + \frac{1}{2}(\alpha - \alpha^{p^{k+s}})x^{p^k}$$

$$= \begin{cases} \frac{1}{2}(\alpha + \alpha^p)x + \frac{1}{2}(\alpha - \alpha^p)x^{p^k} & \text{if } k + s \text{ is odd,} \\ \alpha x & \text{if } k \text{ and } s \text{ are even.} \end{cases}$$

Hence, for $k + s$ odd

$$L^{-1}(x \star \alpha) \star y = \frac{1}{2}\Big((\alpha + \alpha^p)x + \frac{1}{2}(\alpha - \alpha^p)x^{p^k}\Big) \star y$$

$$= \frac{1}{2}\big(b^{p^s+1}((\alpha + \alpha^p)xy^{p^s} + (\alpha + \alpha^p)x^{p^s}y$$

$$+ (\alpha - \alpha^p)x^{p^k}y^{p^s} + (\alpha - \alpha^p)^{p^s}x^{p^{k+s}}y)$$

$$- b^{p^k(p^s+1)}((\alpha + \alpha^p)x^{p^k}y^{p^{k+s}} + (\alpha + \alpha^p)x^{p^{k+s}}y^{p^k}$$

$$+ (\alpha - \alpha^p)^{p^k}xy^{p^{k+s}} + (\alpha - \alpha^p)^{p^{k+s}}x^{p^s}y^{p^k})$$

$$+ \sum_{i=0}^{k-1} c_i((\alpha + \alpha^p)x^{p^i}y^{p^{k+i}} + (\alpha + \alpha^p)x^{p^{k+i}}y^{p^i}$$

$$+ (\alpha - \alpha^p)^{p^i}x^{p^{k+i}}y^{p^{k+i}} + (\alpha - \alpha^p)^{p^{k+i}}x^{p^i}y^{p^i})\big)$$

$$= L^{-1}(y \star \alpha) \star x.$$

If k and s are even

$$L^{-1}(x \star \alpha) \star y = b^{p^s+1}(\alpha x y^{p^s} + \alpha x^{p^s} y)$$

$$- b^{p^k(p^s+1)}(\alpha x^{p^k} y^{p^{k+s}} + \alpha x^{p^{k+s}} y^{p^k})$$

$$+ \sum_{i=0}^{k-1} c_i(\alpha^{p^i} x^{p^i} y^{p^{k+i}} + \alpha^{p^i} x^{p^{k+i}} y^{p^i})$$

$$= L^{-1}(y \star \alpha) \star x.$$

Hence, $L(\mathbb{F}_{p^2})$ is contained in the middle nucleus of the semifield $\mathbb{S}_F$ and, therefore, since nuclei of a semifield are finite fields then the middle nucleus must have a square order. □

Corollary 36 [5] *If k is odd and $b \in \mathbb{F}_{p^k}$ then the PN functions from the first family of the case (iv) define a commutative semifield non-isotopic to Dickson semifields (and therefore they are CCZ-inequivalent to Dickson PN functions).*

Proof The middle nuclei of Dickson semifields have the order p^k (see [11]) which is not a square for k odd. Since the orders of the middle nuclei of isotopic semifields are equal then the commutative semifields defined by the first family of (iv) are non-isotopic to Dickson semifields due to Theorem 36. □

Now we can formulate the main result of this section.

Corollary 37 [5] *If $p \neq 3$ and k is odd then the PN functions*

$$F(x) = x^{p^s+1} - x^{p^{k+s}+p^s} \pm x^{p^k+1}$$

of the first family of the case (iv) are CCZ-inequivalent to all previously known PN functions and define commutative semifields non-isotopic to all previously known semifields (that is, the finite filed, the Alberts commutative twisted fields and Dickson semifields). □

The following proposition gives additional information on the nuclei of semifields defined by the first family of the case (iv).

Proposition 38 [5] *Let F be a PN function of the first family of the case (iv). Then the order of the middle nucleus of $\mathbb{S}_F$ is the $\gcd(s, k)$-th power. Besides, if $c_i = 0$ for i not divisible by s, then the order of the left nucleus of $\mathbb{S}_F$ is the $\gcd(s, k)$-th power as well.*

Proof With notations (6.18–6.19) we are going to prove that equality (6.20) takes place for any $x, y \in \mathbb{F}_{p^{2k}}$ and any $\alpha \in \mathbb{F}_{p^{\gcd(s,k)}}$. Indeed, since $\alpha^{p^s} = \alpha^{p^k} = \alpha$ then

$$L^{-1}(x \star \alpha) = L^{-1}\big(b^{p^s+1}(x\alpha^{p^s} + x^{p^s}\alpha) - b^{p^k(p^s+1)}(x^{p^k}\alpha^{p^{k+s}} + x^{p^{k+s}}\alpha^{p^k})$$

$$+ \sum_{i=0}^{k-1} c_i(x^{p^i}\alpha^{p^{k+i}} + x^{p^{k+i}}\alpha^{p^i})\big)$$

$$= L^{-1}\big(b^{p^s+1}(x\alpha + (x\alpha)^{p^s}) - b^{p^k(p^s+1)}((x\alpha)^{p^k} + (x\alpha)^{p^{k+s}})$$

$$+ \sum_{i=0}^{k-1} c_i((x\alpha)^{p^i} + (x\alpha)^{p^{k+i}}))$$

$$= L^{-1}(L(\alpha x)) = \alpha x. \tag{6.23}$$

Hence,

$$L^{-1}(x \star \alpha) \star y = b^{p^s+1}(\alpha x y^{p^s} + \alpha x^{p^s} y) - b^{p^k(p^s+1)}(\alpha x^{p^k} y^{p^{k+s}} + \alpha x^{p^{k+s}} y^{p^k})$$

$$+ \sum_{i=0}^{k-1} c_i(\alpha^{p^i} x^{p^i} y^{p^{k+i}} + \alpha^{p^i} x^{p^{k+i}} y^{p^i})$$

$$= L^{-1}(y \star \alpha) \star x.$$

Thus, $L(\mathbb{F}_{p^{\gcd(s,k)}})$ is contained in the middle nucleus of the semifield $\mathbb{S}_F$ and, therefore, since nuclei of a semifield are finite fields, the middle nucleus of $\mathbb{S}_F$ has to be a power of $\gcd(s,k)$.

We are going to prove that the equality

$$L^{-1}(x \star \alpha) \star y = L^{-1}(x \star y) \star \alpha \tag{6.24}$$

takes place for any $x, y \in \mathbb{F}_{p^{2k}}$ and any $\alpha \in \mathbb{F}_{p^{\gcd(s,k)}}$. Indeed, since $\alpha^{p^s} = \alpha^{p^k} = \alpha$ then

$$x \star \alpha = b^{p^s+1}(x\alpha^{p^s} + x^{p^s}\alpha) - b^{p^k(p^s+1)}(x^{p^k}\alpha^{p^{k+s}} + x^{p^{k+s}}\alpha^{p^k})$$

$$+ \sum_{i=0}^{k-1} c_{is}(x^{p^{is}}\alpha^{p^{k+is}} + x^{p^{k+is}}\alpha^{p^{is}})$$

$$= b^{p^s+1}(x\alpha + x^{p^s}\alpha) - b^{p^k(p^s+1)}(x^{p^k}\alpha + x^{p^{k+s}}\alpha)$$

$$+ \sum_{i=0}^{k-1} c_{is}(x^{p^{is}}\alpha + x^{p^{k+is}}\alpha)$$

$$= \alpha L(x).$$

Hence,

$$L^{-1}(x \star y) \star \alpha = \alpha L(L^{-1}(x \star y)) = \alpha(x \star y)$$

and using (6.23) we get

$$L^{-1}(x \star \alpha) \star y = (\alpha x) \star y$$

$$= b^{p^s+1}(\alpha x y^{p^s} + \alpha x^{p^s} y) - b^{p^k(p^s+1)}(\alpha x^{p^k} y^{p^{k+s}} + \alpha x^{p^{k+s}} y^{p^k})$$

$$+ \sum_{i=0}^{k-1} c_{is}(\alpha x^{p^{is}} y^{p^{k+is}} + \alpha x^{p^{k+is}} y^{p^{is}})$$

$$= \alpha(x \star y).$$

This proves equality (6.24). Thus, $L(\mathbb{F}_{p^{\gcd(s,k)}})$ is contained in the left nucleus of the semifield $\mathbb{S}_F$ and, therefore, the left nucleus of $\mathbb{S}_F$ is a power of $\gcd(s,k)$. □

Similar results can be proven also for semifields of the second family of (iv) and semifields of (v).

Proposition 39 [5] *Let F be a PN function of the second family of (iv). Then the order of the middle nucleus of the commutative semifield defined by F is divisible by $\gcd(s,k)$.*

Proof For any $x, y \in \mathbb{F}_{p^{2k}}$ we denote

$$
\begin{aligned}
x \star y &= F(x+y) - F(x) - F(y) \\
&= b(xy^{p^s} + x^{p^s}y) + b^{p^k}(x^{p^k}y^{p^{k+s}} + x^{p^{k+s}}y^{p^k}) \\
&\quad + c(xy^{p^k} + x^{p^k}y) + \sum_{i=0}^{k-1} r_i(x^{p^i}y^{p^{k+i}} + x^{p^{k+i}}y^{p^i}).
\end{aligned}
\tag{6.25}
$$

and

$$
L(x) = 1 \star x = b(x + x^{p^s}) + b^{p^k}(x^{p^k} + x^{p^{k+s}}) + c(x + x^{p^k}) + \sum_{i=0}^{k-1} r_i(x^{p^i} + x^{p^{k+i}}).
\tag{6.26}
$$

Then the multiplication $\circ$ of the commutative semifield $\mathbb{S}_F$ defined by F is

$$
x \circ y = L^{-1}(x) \star L^{-1}(y),
\tag{6.27}
$$

for any $x, y \in \mathbb{F}_{p^{2k}}$.

We are going to prove that for any $x, y \in \mathbb{F}_{p^{2k}}$ and any $\alpha \in \mathbb{F}_{p^{\gcd(s,k)}}$

$$
(x \circ L(\alpha)) \circ y = (y \circ L(\alpha)) \circ x,
$$

or, since L is a permutation then, equivalently, we need to prove that

$$
L^{-1}(x \star \alpha) \star y = L^{-1}(y \star \alpha) \star x.
\tag{6.28}
$$

Indeed, since $\alpha^{p^s} = \alpha^{p^k} = \alpha$ then

$$
\begin{aligned}
L^{-1}(x \star \alpha) &= L^{-1}\big(b(x\alpha^{p^s} + x^{p^s}\alpha) + b^{p^k}(x^{p^k}\alpha^{p^{k+s}} + x^{p^{k+s}}\alpha^{p^k}) \\
&\quad + c(x\alpha^{p^k} + x^{p^k}\alpha) + \sum_{i=0}^{k-1} r_i(x^{p^i}\alpha^{p^{k+i}} + x^{p^{k+i}}\alpha^{p^i})\big) \\
&= L^{-1}\big(b(x\alpha + (x\alpha)^{p^s}) + b^{p^k}((x\alpha)^{p^k} + (x\alpha)^{p^{k+s}}) \\
&\quad + c(x\alpha + (x\alpha)^{p^k}) + \sum_{i=0}^{k-1} r_i((x\alpha)^{p^i} + (x\alpha)^{p^{k+i}})\big)
\end{aligned}
$$

$$= L^{-1}(L(\alpha x)) = \alpha x. \tag{6.29}$$

Hence,

$$L^{-1}(x \star \alpha) \star y = (\alpha x) \star y = b(\alpha x y^{p^s} + \alpha x^{p^s} y) + b^{p^k}(\alpha x^{p^k} y^{p^{k+s}} + \alpha x^{p^{k+s}} y^{p^k})$$

$$+ c(\alpha x y^{p^k} + \alpha x^{p^k} y) + \sum_{i=0}^{k-1} r_i(\alpha^{p^i} x^{p^i} y^{p^{k+i}} + \alpha^{p^i} x^{p^{k+i}} y^{p^i})$$

$$= L^{-1}(y \star \alpha) \star x.$$

Thus, $L(\mathbb{F}_{p^{\gcd(s,k)}})$ is contained in the middle nucleus of the semifield $\mathbb{S}_F$ and, therefore, d has to be divisible by $\gcd(s, k)$. $\square$

Proposition 40 [5] *Let F be a PN function of the second family of (iv) where* $r_i = 0$ *for i not divisible by s. If* p^d *is the order of the left nucleus of the commutative semifield defined by F then d is divisible by* $\gcd(s, k)$.

Proof With notations (6.25–6.27) we are going to prove that equallity (6.24) takes place for any $x, y \in \mathbb{F}_{p^{2k}}$ and any $\alpha \in \mathbb{F}_{p^{\gcd(s,k)}}$. Indeed, since $\alpha^{p^s} = \alpha^{p^k} = \alpha$ then

$$x \star \alpha = b(x\alpha^{p^s} + x^{p^s}\alpha) + b^{p^k}(x^{p^k}\alpha^{p^{k+s}} + x^{p^{k+s}}\alpha^{p^k})$$

$$+ c(x\alpha^{p^k} + x^{p^k}\alpha) + \sum_{i=0}^{k-1} r_{is}(x^{p^{is}}\alpha^{p^{k+is}} + x^{p^{k+is}}\alpha^{p^{is}})$$

$$= b(x\alpha + x^{p^s}\alpha) + b^{p^k}(x^{p^k}\alpha + x^{p^{k+s}}\alpha)$$

$$+ c(x\alpha + x^{p^k}\alpha) + \sum_{i=0}^{k-1} r_{is}(x^{p^{is}}\alpha + x^{p^{k+is}}\alpha)$$

$$= \alpha L(x).$$

Hence,

$$L^{-1}(x \star y) \star \alpha = \alpha L(L^{-1}(x \star y)) = \alpha(x \star y)$$

and using (6.29) we get

$$L^{-1}(x \star \alpha) \star y = (\alpha x) \star y$$

$$= b(\alpha x y^{p^s} + \alpha x^{p^s} y) + b^{p^k}(\alpha x^{p^k} y^{p^{k+s}} + \alpha x^{p^{k+s}} y^{p^k})$$

$$+ c(\alpha x y^{p^k} + \alpha x^{p^k} y) + \sum_{i=0}^{k-1} r_{is}(\alpha x^{p^{is}} y^{p^{k+is}} + \alpha x^{p^{k+is}} y^{p^{is}})$$

$$= \alpha(x \star y).$$

This proves equality (6.24). Thus, $L(\mathbb{F}_{p^{\gcd(s,k)}})$ is contained in the left nucleus of the semifield $\mathbb{S}_F$ and, therefore, d has to be divisible by $\gcd(s, k)$. $\square$

Proposition 41 [5] *Let F be a PN function of the family (v). Then the orders of the middle and left nuclei of the commutative semifield defined by F are the $\gcd(s,t)$-th powers.*

Proof For any $x, y \in \mathbb{F}_{p^{3t}}$ we denote

$$x \star y = F(x+y) - F(x) - F(y)$$
$$= xy^{p^s} + x^{p^s}y - a^{p^t-1}(x^{p^t}y^{p^{2t+s}} + x^{p^{2t+s}}y^{p^t}). \tag{6.30}$$

and

$$L(x) = 1 \star x = x + x^{p^s} - a^{p^t-1}(x^{p^t} + x^{p^{2t+s}}). \tag{6.31}$$

Then the multiplication $\circ$ of the commutative semifield $\mathbb{S}_F$ defined by F is

$$x \circ y = L^{-1}(x) \star L^{-1}(y), \tag{6.32}$$

for any $x, y \in \mathbb{F}_{p^{3t}}$.

We are going to prove that for any $x, y \in \mathbb{F}_{p^{3t}}$ and any $\alpha \in \mathbb{F}_{p^{\gcd(s,t)}}$

$$(x \circ L(\alpha)) \circ y = (y \circ L(\alpha)) \circ x,$$
$$(x \circ L(\alpha)) \circ y = (x \circ y) \circ L(\alpha),$$

or, since L is a permutation

then, equivalently, we need to prove that

$$L^{-1}(x \star \alpha) \star y = L^{-1}(y \star \alpha) \star x, \tag{6.33}$$
$$L^{-1}(x \star \alpha) \star y = L^{-1}(x \star y) \star \alpha. \tag{6.34}$$

Since $\alpha^{p^s} = \alpha^{p^t} = \alpha$ then

$$x \star \alpha = x\alpha^{p^s} + x^{p^s}\alpha - a^{p^t-1}(x^{p^t}\alpha^{p^{2t+s}} + x^{p^{2t+s}}\alpha^{p^t})$$
$$= x\alpha + (x\alpha)^{p^s} - a^{p^t-1}((x\alpha)^{p^t} + (x\alpha)^{p^{2t+s}})$$
$$= L(\alpha x) = \alpha L(x).$$

Thus,

$$L^{-1}(x \star \alpha) = L^{-1}(L(\alpha x)) = \alpha x,$$
$$L^{-1}(x \star y) \star \alpha = \alpha L(L^{-1}(x \star y)) = \alpha(x \star y),$$

and therefore

$$L^{-1}(x \star \alpha) \star y = (\alpha x) \star y = \alpha x y^{p^s} + \alpha x^{p^s}y - a^{p^t-1}(\alpha x^{p^t}y^{p^{2t+s}} + \alpha x^{p^{2t+s}}y^{p^t})$$
$$= \alpha(x \star y) = L^{-1}(y \star \alpha) \star x = L^{-1}(x \star y) \star \alpha,$$

which proves equalities (6.33) and (6.34).

Hence, $L(\mathbb{F}_{p^{\gcd(s,t)}})$ is contained in the left and middle nuclei of the semifield $\mathbb{S}_F$. Therefore, if p^{d_l} and p^{d_m} are the orders of the left and middle nuclei of $\mathbb{S}_F$, repectively, then d_l and d_m are divisible by $\gcd(s,t)$. $\square$

6.5 Generalization of Family (vii) of Planar Functions

As previosly mentioned, under some condition on n, Coulter and Henderson proved in [7] that commutative presemifields of order p^n are isotopic if and only if they are strongly isotopic. However, there are cases when isotopic commutative presemifields define CCZ-inequivalent quadratic PN functions, as shown in [10] by using function (vii) with parameters $p = 3$ and $k = 1$. Below we show that this example is generalizable for any odd prime p and any positive integer k. In particular, we extend the family of functions F constructed in [1, 8] to the family (vii) of functions F' below with larger CCZ-equivalence class. Let

$$F(x) = x^2 + x^{2p^m} + \sum_{i=0}^{k}(-1)^i x^{p^{2i}(p^2+1)} + \sum_{j=0}^{k-1}(-1)^{k+j} x^{p^{2j+1}(p^2+1)}$$

$$- \Big(\sum_{i=0}^{k}(-1)^i x^{p^{2i}(p^2+1)} + \sum_{j=0}^{k-1}(-1)^{k+j} x^{p^{2j+1}(p^2+1)}\Big)^{p^m};$$

be a function over $\mathbb{F}_{p^{2m}}$ with $m = 2k + 1$. Then according to [1, 8] it is PN. Let "$\star$" denotes the operation

$$x \star y = F(x + y) - F(x) - F(y)$$

for all $x, y \in \mathbb{F}_{p^n}$. Then we know that the triple $\mathbb{S} = (\mathbb{F}_{p^n}, +, \star)$ is a commutative presemifield. As we can see below it is a semifield with the identity $e \in \mathbb{F}_p$ defined by the condition

$$4e \bmod p = 1.$$

Indeed, for any $a \in \mathbb{F}_{p^2}^*$ we have

$$x \star a = F(x + a) - F(x) - F(a)$$

$$= 2ax + 2a^p x^{p^m} + \sum_{i=0}^{k}(-1)^i (ax^{p^{2(i+1)}} + ax^{p^{2i}})$$

$$+ \sum_{j=0}^{k-1}(-1)^{k+j}(a^p x^{p^{2(j+1)+1}} + a^p x^{p^{2j+1}})$$

$$- \Big(\sum_{i=0}^{k}(-1)^i (ax^{p^{2(i+1)}} + ax^{p^{2i}})$$

$$+ \sum_{j=0}^{k-1} (-1)^{k+j} \left(a^p x^{p^{2(j+1)+1}} + a^p x^{p^{2j+1}} \right) \Big)^{p^m}$$

$$= 2ax + 2a^p x^{p^m} + \left(ax + (-1)^k a x^{p^{m+1}} \right) + \left((-1)^{k+1} a^p x^p - a^p x^{p^m} \right)$$

$$- \left(\left(ax + (-1)^k a x^{p^{m+1}} \right) + \left((-1)^k a^p x^p - a^p x^{p^m} \right) \right)^{p^m}$$

$$= 4ax.$$

We can see that $L(x) = x * a$ is a linear permutation, and, therefore, the operation

$$x \circ y = (x * a) * y$$

defines a comutative semifield $\mathbb{S}' = (\mathbb{F}_{p^n}, +, \circ)$ isotopic to $\mathbb{S}$ and with identity $e' = a^{-1}d$, where $d \in \mathbb{F}_p$ is defined from

$$16d \bmod p = 1.$$

Then the function $F'(x) = x \circ x$ is planar and

$$F'(x) = (x \star a) \star x = (4ax) \star x$$

$$= 4 \Big(2ax^2 + 2a^p x^{2p^m} + 2a \sum_{i=0}^{k} (-1)^i x^{p^{2i}(p^2+1)}$$

$$+ 2a^p \sum_{j=0}^{k-1} (-1)^{k+j} x^{p^{2j+1}(p^2+1)}$$

$$- \Big(2a \sum_{i=0}^{k} (-1)^i x^{p^{2i}(p^2+1)} + 2a^p \sum_{j=0}^{k-1} (-1)^{k+j} x^{p^{2j+1}(p^2+1)} \Big)^{p^m} \Big)$$

$$= 8a \Big(x^2 + a^{p-1} x^{2p^m} + \sum_{i=0}^{k} (-1)^i x^{p^{2i}(p^2+1)}$$

$$+ a^{p-1} \sum_{j=0}^{k-1} (-1)^{k+j} x^{p^{2j+1}(p^2+1)}$$

$$- \Big(a^{1-p} \sum_{i=0}^{k} (-1)^i x^{p^{2i}(p^2+1)} + \sum_{j=0}^{k-1} (-1)^{k+j} x^{p^{2j+1}(p^2+1)} \Big)^{p^m} \Big).$$

Obviously, $\mathbb{S}$ and $\mathbb{S}'$ are isotopic by construction. According to the theorem below, $\mathbb{S}$ and $\mathbb{S}'$ can be potentially non-strongly isotopic if $a \in N_m(\mathbb{S}) \setminus N(\mathbb{S})$.

Theorem 37 [7] *Let* $\mathbb{S}_1 = (\mathbb{F}_{p^n}, +, \star)$ *and* $\mathbb{S}_2 = (\mathbb{F}_{p^n}, +, \circ)$ *be isotopic commutative semifields. Then there exists an isotopism* (M, N, L) *between* $\mathbb{S}_1$ *and* $\mathbb{S}_2$ *such that either* *(i)* $M = N$, *or*

(ii) $M(X) = \alpha \star N(X)$ *where* $\alpha \in N_m(\mathbb{S}_1) \setminus N(\mathbb{S}_1)$.

It can be easily checked that

$$(x \star a) \star y = (4ax) \star y = (4ay) \star x = (y \star a) \star x$$

for any $x, y \in \mathbb{F}_{p^{2m}}$. That is, $\mathbb{F}_{p^2}$ is a subset of $N_m(\mathbb{S})$. On the other hand, $a \star (x \star y) = 4a(x \star y)$, and if $(4ax) \star y = 4a(x \star y)$ for any $x, y \in \mathbb{F}_{p^{2m}}$, then the coefficient of the monomial $x^{p^m} y^{p^m}$ would be the same in $(4ax) \star y$ and $4a(x \star y)$, while it is $8a^p$ in the first one and $8a$ in the second. Hence, $a \in N_m(\mathbb{S}) \setminus N(\mathbb{S})$ for $a \in \mathbb{F}_{p^2} \setminus \mathbb{F}_p$. According to [10], if $p = 3$ and $m = 3$ then $\mathbb{S}$ and $\mathbb{S}'$ are not strongly isotopic. Hence the function F' is a generalization of the function F: it coincides with F for $a \in \mathbb{F}_p^*$ and, in general, it is CCZ-inequivalent to F for $a \in \mathbb{F}_{p^2} \setminus \mathbb{F}_p$. We can formulate now the following theorem.

Theorem 38 [4] *Let k be a positive integer, $m = 2k + 1$ and $a \in \mathbb{F}_{p^2}^*$. Then the function*

$$F'(x) = 8a\left(x^2 + a^{p-1}x^{2p^m} + \sum_{i=0}^{k}(-1)^i x^{p^{2i}(p^2+1)}\right.$$

$$+ a^{p-1}\sum_{j=0}^{k-1}(-1)^{k+j}x^{p^{2j+1}(p^2+1)}$$

$$\left. - \left(a^{1-p}\sum_{i=0}^{k}(-1)^i x^{p^{2i}(p^2+1)} + \sum_{j=0}^{k-1}(-1)^{k+j}x^{p^{2j+1}(p^2+1)}\right)^{p^m}\right).$$

is planar over $\mathbb{F}_{p^{2m}}$. For different $a \in \mathbb{F}_{p^2}^*$ the functions F' define isotopic commutative semifields. Two functions F' defined with $a \in \mathbb{F}_{p^2} \setminus \mathbb{F}_p$ and $a \in \mathbb{F}_p^*$, respectively, are not CCZ-equivalent for $m = 3$ and the corresponding commutative semifields are not strongly isotopic. When $a = 1$ the function F is the one constructed in [1, 8].

References

1. J. Bierbrauer. Commutative semifields from projection mappings. *Designs, Codes and Cryptography*, 61(2), pp. 187–196, 2011.
2. C. Bracken, E. Byrne, N. Markin, G. McGuire. New families of quadratic almost perfect nonlinear trinomials and multinomials. Finite Fields and Their Applications **14**(3), pp. 703–714, 2008.
3. L. Budaghyan and T. Helleseth. New perfect nonlinear multinomials over $F_{p^{2k}}$ for any odd prime p. *Proceedings of the International Conference on Sequences and Their Applications SETA 2008*, Lecture Notes in Computer Science 5203, pp. 403–414, Lexington, USA, Sep. 2008.

4. L. Budaghyan and T. Helleseth. On Isotopisms of Commutative Presemifields and CCZ-Equivalence of Functions. Special Issue on Cryptography of *International Journal of Foundations of Computer Science*, v. 22(6), pp. 1243–1258, 2011. Preprint at http://eprint.iacr.org/2010/507

5. L. Budaghyan and T. Helleseth. New commutative semifields defined by new PN multinomials. *Cryptography and Communications: Discrete Structures, Boolean Functions and Sequences*, v. 3(1), pp. 1–16, 2011.

6. L. Budaghyan, C. Carlet, G. Leander. Two classes of quadratic APN binomials inequivalent to power functions. *IEEE Trans. Inform. Theory*, 54(9), pp. 4218–4229, 2008.

7. R. S. Coulter and M. Henderson. Commutative presemifields and semifields. *Advances in Math.* 217, pp. 282–304, 2008.

8. G. Lunardon, G. Marino, O. Polverion, R. Trombetti. Symplectic spreads and quadric Veroneseans. Manuscript, 2009.

9. G. J. Ness. Correlation of sequences of different lengths and related topics. PhD dissertation. University of Bergen, Norway, 2007.

10. Y. Zhou. A note on the isotopism of commutative semifields. Preprint, 2010.

11. L. E. Dickson. Linear algebras with associativity not assumed. *Duke Math. J.* 1(2), pp. 113–125, 1935.

Printed in the United States
By Bookmasters